高等教育应用型重点专业规划教材

New English Course in Electromechanics

新编机电英语

丛书主编：冯光华

主编：胡　杨　关荆晶　陈　斯

副主编：贾　文　徐志鹏

天津大学出版社
TIANJIN UNIVERSITY PRESS

图书在版编目（ＣＩＰ）数据

新编机电英语 / 胡杨，关荆晶，陈斯主编． —天津：天津大学出版社，2018.5
高等教育应用型重点专业规划教材 / 冯光华主编（2023.8 重印）

ISBN 978-7-5618-6130-1

Ⅰ．①新… Ⅱ．①胡… ②关… ③陈… Ⅲ．①机电工程－英语－高等
学校－教材 Ⅳ．① TH

中国版本图书馆 CIP 数据核字（2018）第 103939 号

出版发行	天津大学出版社
地　　址	天津市卫津路 92 号天津大学内（邮编：300072）
电　　话	发行部：022-27403647
网　　址	publish.tju.edu.cn
印　　刷	北京盛通印刷股份有限公司
经　　销	全国各地新华书店
开　　本	787mm×1092mm　1/16
印　　张	7.25
字　　数	201 千
版　　次	2018 年 5 月第 1 版
印　　次	2023 年 8 月第 2 次
定　　价	35.00 元

丛书编委会

总主编：冯光华

编　委：（排名不分先后）

《新编机电英语》编委会

前言

　　《机电英语》作为校内教材使用的时候，得到了广大读者的肯定，并受到了广大学生的好评。与此同时，我们也收到了武汉工程科技学院机电学院相关专业师生提出的不少宝贵意见。因此，根据编者的教学实践以及与机电行业相关专家合作积累的经验，正式修订并出版了这本《新编机电英语》。

　　本书主要根据武汉工程科技学院机电学院各专业的英语口语教材和机电英语选修课教材编写而成。在武汉工程科技学院多年教学实践的基础上，编者对上述内容进行了进一步的修改、充实、提高，改编成本书。《新编机电英语》教材遵循"工学结合，能力为本"的教学理念和"实用为主，够用为度"的教学原则编写，以期培养和提高学生实际运用英语语言的能力。全书共8个单元，每个单元分为专业词汇、实用对话、阅读、知识拓展、综合练习等五个部分。其中实用对话部分选择的是职场中的涉外英语内容，设置真实的语境，培养学生的英语交流能力；阅读部分选用与电机机械、机电建模、机电一体化、电子通信、信息技术、云计算、系统工程专业相关的文章。《新编机电英语》内容新颖、通俗、实用，既可以提高学生的语言技能，又有利于培养学生的职业素质与技能。全书图文并茂，集职业性、实用性、适时性和趣味性于一体。这8个单元的内容既自成体系，又互相关联，难度适中，为配合教学，还编有一定量的练习，供学生在课堂内外使用。建议学生在使用时多精读课本，以提高学习效果。

　　本教材注重英语听说口语练习和专业英语阅读练习，以下主要包含形式。

　　1．口语练习，训练形式多样，既有对每章主题的相关内容进行双人交流或进行拓展性小组讨论，也有就图表信息或提示信息进行个人陈述等。口语练习的设计旨在培养学生运用英语表达相关专业知识的能力，做到信息输入和信息输出同步进行。

　　2．专业英语阅读练习，阅读部分选用与电机机械、机电建模、机电一体化、电子通信、信息技术、云计算、系统工程专业相关的文章，每个单元从课文到练习上的设计上由浅入深，既突出了基础知识也强调实际应用，既突出专业特色又能充分体现英语教学的规律，达到语言技能与职业知识技能的整合。

　　《新编机电英语》适合作为应用技术型高校机电专业的专业英语教材，供应用技术型高校师生使用。

编者

2018 年 1 月

目 录

Unit 1 Electromechanics

Learning Objectives

After completing this unit, you will be able to do the following:

1. Grasp the main idea and the structure of the text;
2. Master the key language points and grammatical structure in the text;
3. Understand the basic concepts in electromechanics;
4. Conduct a series of reading, listening, speaking and writing activities related to the theme of the unit.

Outline

The followings are the main sections of this unit:

1. Warm-up Activity;
2. Text A;
3. Situational Conversation;
4. Reading Comprehension;
5. Translation Skills;
6. Exercises.

Technical Terms

In this unit, you will learn some technical terms in electromechanics listed below:

- electrical engineering;
- mechanical engineering;
- electric typewriters;
- digital computers;
- relays.

Vocabulary

Listed below are some words appearing in this unit that you should make part of your vocabulary:

- microcontroller;
- traffic lights;
- washing machines;
- telegraphy;
- voltage;
- encompass;
- integration;
- indefinitely;
- selectric.

👉 Looking Ahead

The purpose of Electro-mechanical Modelling is to model and simulate an electro-mechanical system, so that its physical parameters can be examined before the actual system is built. Parameter estimation and physical realization of the overall system are the major design objectives of electro-mechanical modelling. Theory driven mathematical models can be used for or applied to other systems to judge the performance of the joint system as a whole.

The modelling of pure mechanical systems is mainly based on the Lagrangian which is a function of the generalized coordinates and the associated velocities. If all forces are derivable from a potential, then the time behavior of the dynamic systems is completely determined. For simple mechanical systems, the Lagrangian is defined as the difference of the kinetic energy and the potential energy.

In consequence, we have quantities (kinetic and potential energy, generalized forces) which determine the mechanical part and quantities (co-energy, powers) for the description of the electrical part. This offers a combination of the mechanical and electrical parts by means of an energy approach. As a result, an extended Lagrangian format is produced.

Introduction

The Lagrangian System is mainly used in the process of modelling of pure mechanical systems.

Core Contents

1. The discussion of the definition of electromechanics.

2. Discussion of the automated telephone exchanges that were widely used in the early days.

◈ Warm-up Activity

What was widely used in early automated telephone exchanges?

The solenoid valve

A steam turbine used to provide electric power

▣ Part One: Text A History of Electromechanics

In engineering, electromechanics combines electrical and mechanical processes and procedures drawn from electrical engineering and mechanical engineering. Electrical engineering in this context also encompasses electronics engineering.

Devices which carry out electrical operations by using moving parts are known as electromechanics. Strictly speaking, a manually operated switch is an electromechanical component, but the term is usually understood to refer to devices which involve an electrical signal to create mechanical movement, or mechanical movement to create an electric signal, often involving electromagnetic principles such as in relays, which allow a voltage or current to control other, oftentimes isolated circuit voltage or current by mechanically switching sets of contacts, and solenoids, by which the voltage can actuate a moving linkage as in solenoid valves. Piezoelectric devices are electromechanical, but do not use electromagnetic principles. Piezoelectric devices can create sound or vibration from an electrical signal or create an electrical signal from sound or mechanical vibration.

Before the development of modern electronics, electromechanical devices were widely used in complicated systems and subsystems, including electric typewriters, teleprinters, very early television systems, and the very early electromechanical digital computers.

Relays originated with telegraphy as electromechanical devices used to regenerate telegraph signals. In 1885, Michael Pupin at Columbia University taught mathematical physics and electromechanics until 1931.

The Strowger switch, the Panel switch, and similar ones were widely used in early automated telephone exchanges. Crossbar switches were first widely installed in the middle 20th century in Sweden, the United States, Canada, and the Great Britain, and these quickly spread to the rest of the world—especially Japan. The electromechanical television systems of the late 19th century were less successful.

Electric typewriters developed, up to the 1980s, as "power-assisted typewriters". They contained a single electrical component, the motor. The keystroke used to move a typebar directly, but now it engaged mechanical linkages that directed mechanical power from the motor into the type bar. This was also true of the later IBM Selectric. At Bell Labs, in the 1940s, the Bell Model V computer was developed. It was an electromechanical relay-based device; cycles took seconds. In 1968 electromechanical systems were still under serious consideration for an aircraft flight control computer, until a device based on large scale integration electronics was adopted in the Central Air Data Computer.

At the beginning of the last third of the century, much equipment which for most of the 20th century would have used electromechanical devices for control, has come to use less expensive and more reliable integrated microcontroller circuits containing millions of transistors, and a program to carry out the same task through logic, with electromechanical components only where moving parts, such as mechanical electric actuators, are a requirement. Such chips have replaced most electromechanical devices, because any point in a system which must rely on mechanical movement for proper operation would have mechanical wear and eventually fail. Properly designed electronic circuits without moving parts will continue to operate properly almost indefinitely and are used in most simple feedback control systems, and would appear in huge numbers in everything from traffic lights to washing machines.

As in 2010, approximately 16,400 people work as electro-mechanical technicians in the US, about 1 out of every 9,000 workers. Their median annual wage is about 50% more than the median annual wage over all occupations.

Words and Expressions

electromechanics [ɪˌlektrəʊmɪˈkænɪks]   *n.* 电机机械

procedure	[prə'si:dʒə]	n. 程序，手续；步骤
be drawn from		从……中得到；从……提取
encompass	[ɪn'kʌmpəs; en-]	vt. 包含；包围，环绕；完成
strictly speaking		严格地说；严格来说
voltage	['voltɪdʒ]	n. ［电］电压
current	['kʌr(ə)nt]	n.（水、气、电）流；趋势；涌流
solenoid	['səʊlənɒɪd]	n. ［电］螺线管；螺线形电导管
valve	[vælv]	n. 阀；［解剖］瓣膜；真空管；活门
relay	['ri:leɪ]	n. ［电］继电器
originate with		源于
switch	[swɪtʃ]	n. 开关；转换；鞭子　vi. 转换；抽打；换防
keystroke	['ki:strəʊk]	n. 击键；按键
typebar	['taɪpbɑː]	n. 铅字连动杆
selectric	[si'lektrɪk]	n. 电动打字机
under serious consideration		在认真考虑之下
integration	[ɪntɪ'greɪʃ(ə)n]	n. 集成；综合
microcontroller	[ˌmaɪkrəʊkən'trəʊlə]	n. ［自］微控制器
indefinitely	[ɪn'defɪnɪtlɪ]	adv. 不确定地，无限期地；模糊地，不明确地
feedback	['fi:dbæk]	n. 反馈；成果，资料；回复
in huge numbers		大量的

Exercise 1: Special terms.

1. electrical and mechanical　_____

2. electrical engineering　_____

3. mechanical engineering　_____

4. 电信号　_____

5. 机械联动装置　_____

6. 一个电子元件　_____

Exercise 2: Answer the following questions.

1. What is electromechanics?

2. What is a manually operated switch?

3. What were widely used in early automated telephone exchanges?

4. As in 2010, how many people work as electro-mechanical technicians in the US?

Exercise 3: Define the following terms with information from Text A.

1. electromechanics

2. relay

3. microcontroller

🗐 Part Two : Situational Conversation

At a Chinese Restaurant

A: It's very nice of you to invite me.

B: I'm very glad you could come, Mr. Liu. Would you like to take a seat at the head of the table? It's an informal dinner, please don't stand on ceremony… Mr. Liu, would you like to have some chicken?

A: Thank you. This is my first time to come to a Chinese restaurant. Could you tell me the different features of Chinese food?

B: Generally speaking, Cantonese food is a bit light; Shanghai food is rather oily; and Hunan dishes are very spicy, with a strong and hot taste.

A: Chinese dishes are exquisitely prepared, delicious, and very palatable. They are very good in colour, flavor and taste.

B: Mr. liu, would you care for another help?

A: No more, thank you. I'm quite full.

B: Did you enjoy the meal?

A: It was the most delicious dinner I've had for a long time. It was such a rich dinner.

B: I'm so glad you like it.

A: Thank you very much for your hospitality.

Notes:

At a Chinese Restaurant
在中餐馆

B: I'm very glad you could come, Mr. Liu. Would you like to take a seat at the head of the table? It's an informal dinner, please don't stand on ceremony… Mr. Liu, would you like to have some chicken?

我很高兴您能来，刘先生，请您坐在上席吧？这只是个便饭，请别客气。刘先生，您想吃点鸡肉吗？

A: Could you tell me the different features of Chinese food?

您能告诉我中国菜的不同特点吗？

B: Generally speaking, Cantonese food is a bit light; Shanghai food is rather oily; and Hunan dishes are very spicy, having a strong and hot taste.

一般而言，广东菜较为清淡；上海菜比较油腻；湖南菜非常辣，热辣十足！

A: They are very good in colour, flavor and taste.

它们色香味俱全。

A: It was the most delicious dinner I've had for a long time. It's such a rich dinner.

这是我吃过最美味的宴席了，太丰盛了。

Exercise 1: Sentence patterns.

1. It's very nice of you to invite me.

2. Chinese dishes are exquisitely prepared, delicious, and very palatable. They are very good in colour, flavor and taste.

3. Could you tell me the different features of Chinese food?

4. It's an informal dinner.

5 I'm quite full.

Exercise 2: Complete the following dialogue in English.

A: Now, you are boarding the plane. We're sorry that we haven't done much to help you when you are in China.

B: I appreciate what you have done for me. Everything I have seen here has left a deep impression on me. I really don't know how to express my thanks to you.

A: ____1____

我们很乐意给你提供帮助。

B: ____2____

请代我向张先生和其他的朋友们转达我的谢意好吗？

A: I'd like to. I'm sure your visit will help to promote the friendship and understanding between us. Welcome to China again.

我一定转达。我坚信你的来访将促进我们双方的友谊和了解，欢迎你再访中国。

B: ____3____

当然，我会的。好吧，是告别的时候了，飞机就要起飞了，希望你将来有机会来美国。

A: ____4____

谢谢。如果有机会，我会去的。再见，一路平安！

B: Goodbye!

⚑ Part Three: Reading Comprehension

Questions 1 to 5 are based on the following passage.

For years, high school students have received identical textbooks as their classmates. Even as students have different learning styles and abilities, they are force-fed the same materials, "Imagine a digital textbook my book because I'm a different person and learn differently, which is different from yours," said Richard Baraniuk, founder of OpenStax.

OpenStax will spend two years developing the personalized books and then test them on Houston-area students. The books will also go through a review and evaluation process similar to traditional textbooks. Baraniuk expects 60 people to review each book before the publication to ensure its quality.

The idea is to make learning easier, so students can go on to more successful careers and lives. Baraniuk isn't just reproducing physical textbooks on digital devices, a mistake e-book publishers have made. He's seriously rethinking that the educational experience should be in a world of digital tools. To do this means involving individuals with skills traditionally left out of the textbook business. Baraniuk is currently hiring cognitive scientists and machinelearning experts. Baraniuk wants to use the tactics (策略) of Google, Netflix and Amazon to deliver a personalized experience. These web services all rely on complex algorithmsto (算法) automatically adjust their offerings for customers.

Just as Netflix recommends different movies based on your preference and viewing history, a textbook might present materials at different pace. Thanks to machine learning the textbook—which will be stored on a range of digital devices—will automatically adjust itself. As a student learns about a topic, he or she could be interrupted by brief quizzes that evaluate whether he or she masters the area. Depending on how the student does, the subject could be reinforced with more materials. Or a teacher could be automatically emailed that the student is struggling with a certain concept and could use some one-to-one attention.

This personalized learning experience is possible thanks to the wealth of data that a digital textbook can track. The data can be used to better track students' progress during a course. Parents and teachers can monitor students' development and provide more proper assistance in time. With personalized learning methods, our students' talents will be better developed.

1. What do we learn about personalized books?

A) Their quality will be ensured since they are developed by OpenStax.

B) They will be examined and judged before being published.

C) They will overlook different learning styles and abilities.

D) They will be much similar to traditional textbooks.

2. In which aspect have e-book publishers done incorrectly?

A) They have only put emphasis on learning experience.

B) They have made it difficult to access to e-books.

C) They have made it rather boring and inconvenient to learn.

D) They have just produced an electronic copy of print textbooks.

3. What does Richard Baraniuk mean by "the educational experience should be in a world of digital tools" (Line 3-4，Para. 3)?

A) Education should employ the machine to improve learning.

B) Education should involve traditional textbooks in the digital world.

C) Education should include obtaining skills by adapting machine learning.

D) Education should reproduce traditional textbooks on the Web services.

4. Personalized textbooks are beneficial to students because _____.

A) they store the fixed material on different digital machines.

B) they quiz the students to make them more confident.

C) they automatically present movies based on students' preference.

D) they automatically match learning material to students' needs.

5. Personalized learning experience may become possible owing to _____.

A) a great amount of digital equipment

B) the students' continuous progress

C) a great amount of digital information

D) parents' and teachers' constant attention

Part Four: Translation Skills

翻译技巧：增译法

常用的翻译技巧有增译法、省译法、重复法、转换法、拆句和合并法、正译法、反译法、倒置法、包孕法等。本章着重介绍增译法。

增译法指根据英汉两种语言不同的思维方式、语言习惯和表达方式，在翻译时增添一些单词、短语或句子，以便更准确地表达出原文所包含的意思。这种方式多半用在汉译英中。汉语中无主句的情况较多，而英语句子中则一般都要有主语，所以在翻译汉语无主句的时候，除了少数情况可用英语无主句、被动语态或"There be..."结构来翻译之外，一般都要根据语境补出主语，使句子更完整。

英汉两种语言在名词、代词、连词、介词和冠词的使用方法上也存在很大差别。英语中代词使用频率较高，凡说到人的器官和归某人所有的或与某人有关的事物时，必须在前面加上物主代词。因此，在汉译英时需要增补物主代词，而在英译汉时又需要根据情况适当删减。英语中词与词、词组与词组以及句子与句子的逻辑关系一般用连词来表示，而汉语则往往通过上下文和语序来表示这种关系。因此，在汉译英时常常需要增补连词。此外，英语句子也离不开介词和冠词。

在汉译英时还要注意增补一些原文中暗含而没有明言的词语以及一些概括性、注释性的词语，以确保译文意思的完整。总之，通过增译一是可以保证译文语法结构的完整，二是可以保证译文意思的明确。

例1： In the evening, after the banquet, the concert and table tennis exhibition, he would work on the drafting of the final communiqué.

翻译：晚上在参加宴会、出席音乐会、观看乒乓球表演之后，他还得起草最后公报。（增译动词）

详解：根据译文意思的需要，可以在名词前增加动词。比如把例1中的 after the banquet, the concert and table tennis exhibition 译为"在宴会、音乐会、乒乓球表演之后"，意思就不够明确，而如果在名词之前增加原文中虽无其词却有其意的动词，译为"在参加宴会、出席音乐会、观看乒乓球表演之后"，形成3个动宾词组，意思就明确了，读起来也较通顺自然，符合汉语习惯。

例2： O, Tom Canty, born in rags and dirt and misery, what sight is this!

翻译：哦，汤姆·康第，生在破烂、肮脏和苦难中，现在这番景象却是多么煊赫啊！（增译形容词）

详解：根据原著，汤姆·康第本是个贫儿，穿上王子服装以后，被人认为真的是王子，就显得特别煊赫。原文虽无"煊赫"的字眼，但含有此意，所以翻译时应增加上去。为了满足意思或修辞上的需要，对于一些英语句子中的名词，在翻译成汉语时可以增加一些适当的形容词，从而使语句更为通顺。

例3： In April，there was the "ping" heard around the world. In July, the ping "ponged".

翻译：四月里，全世界听到中国"乒"的一声把球打了出去；到了七月，美国"乓"的一声把球打了回来。（增译背景词语）

详解：翻译有时需要根据上下文及背景情况增加词语，比如例3如果直译就是"四月里全世界听到"乓"的一声；七月里，这"乓"声却"乓"了一下"，读者就会不知所云。同样的增译还可以用在中文成语和谚语的翻译上。例如"三个臭皮匠，顶一个诸葛亮。"增译注释性词语，为"Three cobblers with their wits combined equal Zhuge Liang the mastermind."

Translation Exercise 1

通过使用移动部件进行电气操作的设备被称为机电设备。严格来说，一个手动开关就是一种机电组件，但这一术语通常被理解为涉及一个电信号产生机械运动的设备，或创建一个电信号机械运动。机电这一概念通常指的是涉及电磁继电器等原则，允许电压或电流控制，通常由机械开关的隔离电路电压或使用电流接触和螺线管，电压可来开动一个移动连接电磁阀。光电传感器也属于机电，但没有使用电磁原则。电传感器可以从一个电信号创建声音或振动，也可以从声音或机械振动创建一个电信号。

Translation Exercise 2

投标人须在投标分项报价表后单列、报出买方人员出国参加设计联络会、工厂检验和买方人员在国外接受卖方培训的人均日均费用。以上三项单列的费用，供买方在签订合同时参考，不包括在投标总报价中，目的是为方便买方对最终合同价的组成进行选择和比较，这并不限制买方采用任何一种报价或几种报价组合而签订合同的权力。

Unit 2 Mechanical Engineering

Learning Objectives

After completing this unit, you will be able to do the following:

1. Grasp the main idea and the structure of the text;
2. Master the key language points and grammatical structure in the text;
3. Understand the basic concepts in mechanical engineering;
4. Conduct a series of reading, listening, speaking and writing activities related to the theme of the unit.

Technical Terms

In this unit, you will learn some technical terms in mechanical engineering listed below:

- mechanical engineering;
- mechanical systems;
- engineering disciplines;
- industrial revolution;
- chemical engineering;
- electric motors;
- servo-mechanisms.

Outline

The followings are the main sections of this unit.

1. Warm-up Activity;
2. Text A;
3. Situational Conversation;
4. Reading Comprehension;
5. Translation Skills;
6. Exercises.

Vocabulary

The listed below are some words appearing in this unit that you should make part of your vocabulary:

- manufacturing;
- mechanics;
- kinematics;
- thermodynamics;
- aircraft;
- watercraft;
- robotics;
- incorporate.

☞ Looking Ahead

Mechanical engineering is a discipline that applies the principles of <u>engineering</u>, <u>physics</u>, and <u>materials science</u> for the design, analysis, <u>manufacturing</u>, and maintenance of <u>mechanical systems</u>. It is a branch of engineering that involves the design, production, and operation of <u>machinery</u>. It is one of the oldest and broadest <u>engineering disciplines</u>.

The engineering field requires an understanding of core concepts including <u>mechanics</u>, <u>kinematics</u>, <u>thermodynamics</u>, materials science, <u>structural analysis</u>, and <u>electricity</u>. Mechanical engineers use these core principles along with tools like <u>computer-aided design</u>, and <u>product lifecycle management</u> to design and analyze <u>manufacturing plants</u>, industrial equipment and machinery, <u>heating and cooling systems</u>, <u>transport</u> systems, <u>aircraft</u>, <u>watercraft</u>, <u>robotics</u>, <u>medical devices</u>, <u>weapons</u>, and others.

Introduction

The engineering field requires an understanding of core concepts of mechanics.

Core Contents

1. The discussion of the definition of mechanical engineering.
2. The discussion of the subdisciplines of mechanics.

Warm-up Activity

What does the engineering field require?

Electronic circuit　　　　　　　Complex power systems

 Part One: Text A　Mechanical Engineering

Mechanical engineering is a discipline that applies the principles of engineering, physics, and materials science for the design, analysis, manufacturing, and maintenance of mechanical systems. It is a branch of engineering that involves the design, production, and operation of machinery. It is one of the oldest and broadest engineering disciplines.

The engineering field requires an understanding of core concepts including mechanics, kinematics, thermodynamics, materials science, structural analysis, and electricity. Mechanical engineers use these core principles along with tools like computer-aided design, and product lifecycle management to design and analyze manufacturing plants, industrial equipment and machinery, heating and cooling systems, transport systems, aircraft, watercraft, robotics, medical devices, weapons, etc.

Mechanical engineering emerged as a field during the industrial revolution in Europe in the 18th century; however, its development can be traced back several thousand ago. Mechanical engineering science emerged in the 19th century as a result of developments in the field of physics. The field has continually evolved to incorporate advancement in technology, and mechanical engineers today are pursuing developments in such fields as composites, mechatronics, and nanotechnology. Mechanical engineering overlaps with aerospace engineering, metallurgical engineering, civil engineering, electrical engineering, manufacturing engineering, chemical engineering, and other engineering disciplines to varying amounts. Mechanical engineers may also work in the field of biomedical engineering, specifically with biomechanics, transport phenomena, biomechatronics, bionanotechnology, and modelling of biological systems.

The field of mechanical engineering can be thought as a collection of many mechanical engineering science disciplines. Several of these subdisciplines which are typically taught

新编机电英语

14

at the undergraduate level are listed below, with a brief explanation and the most common application of each. Some of these subdisciplines are unique to mechanical engineering, while others are a combination of mechanical engineering and one or more other disciplines. Most work that a mechanical engineer does uses skills and techniques from several of these subdisciplines, as well as specialized subdisciplines. Specialized subdisciplines, as used in this article, are more likely to be the subject of graduate studies or on-the-job training than undergraduate research. Several specialized subdisciplines are discussed in this section.

Generally speaking, mechanics is the study of forces and their effect upon matter. Typically, engineering mechanics is used to analyze and predict the acceleration and deformation (both elastic and plastic) of objects under known forces (also called loads) or stresses. Subdisciplines of mechanics include:

1. Statics, the study of non-moving bodies under known loads, how forces affect static bodies;

2. Dynamics (or kinetics), the study of how forces affect moving bodies;

3. Mechanics of materials, the study of how different materials deform under various types of stress;

4. Fluid mechanics, the study of how fluids react to forces;

5. Kinematics, the study of the motion of bodies (objects) and systems (groups of objects), while ignoring the forces that cause the motion. Kinematics is often used in the design and analysis of mechanisms;

6. Continuum mechanics, a method of applying mechanics that assumes that objects are continuous (rather than discrete).

Mechanical engineers typically use mechanics in the design or analysis phases of engineering. If the aim of the engineering project is the design of a vehicle, statics might be employed to design the frame of the vehicle, in order to evaluate where the stresses will be most intense. Dynamics might be used when designing the car's engine, to evaluate the forces in the pistons and cams as the engine cycles. Mechanics of materials might be used to choose appropriate materials for the frame and engine. Fluid mechanics might be used to design a ventilation system for the vehicle, or to design the intake system for the engine.

Mechatronics is the combination of mechanics and electronics. It is an interdisciplinary branch of mechanical engineering, electrical engineering and software engineering and it is concerned with integrating electrical and mechanical engineering to create hybrid systems. In this way, machines can be automated through the use of electric motors, servo-mechanisms, and other electrical systems in conjunction with special software. A common example of a mechatronics system is a CD-ROM drive. Mechanical systems open and close the drive, spin the CD and move the laser, while an optical system reads the data on the CD and converts it to bits. Integrated software controls the process and communicates the contents of the CD to the computer.

Robotics is the application of mechatronics to create robots, which are often used in industry to perform tasks that are dangerous, unpleasant, or repetitive. These robots may be of any shape and size, but all are preprogrammed and interact physically with the world. To create a robot, an engineer typically employs kinematics (to determine the robot's range of motion) and mechanics (to determine the stresses within the robot).

Robots are used extensively in industrial engineering. They allow businesses to save money on labor, perform tasks that are either too dangerous or too precise for humans to perform them economically, and to ensure better quality. Many companies employ assembly lines of robots, especially in automotive industries and some factories are so robotized that they can run by themselves. Outside the factory, robots have been employed in bomb disposal, space exploration, and many other fields. Robots are also sold for various residential applications, from recreation to domestic applications.

As the basis of Finite Element Analysis (FEA) or Finite Element Method (FEM) dates back to 1941, this field is not new. But evolution of computers has made FEA/FEM a viable option for analysis of structural problems. Many commercial codes such as ANSYS, Nastran and ABAQUS are widely used in industry for research and design of components. Calculix is an open source and free finite element program. Some 3D modelling and CAD software packages have added FEA modules. Other techniques such as finite difference method (FDM) and finite-volume method (FVM) are employed to solve problems relating heat and mass transfer, fluid flows, fluid surface interaction and etc.

Words and Expressions

discipline	['dɪsɪplɪn]	n. 学科
principle	['prɪnsɪp(ə)l]	n. 原理
analysis	[ə'nælɪsɪs]	n. 分析
manufacturing	[ˌmænjʊ'fæktʃərɪŋ]	n. 制造业；工业
maintenance	['mentənəns]	n. 维护，维修
mechanics	[mɪ'kænɪks]	n. 机械学（用作单数）
kinematics	[ˌkɪnɪ'mætɪks]	n. ［力］运动学；动力学
thermodynamics	[ˌθɜːmə(ʊ)daɪ'næmɪks]	n. ［复数，动词用单数］热力学
computer-aided design		电脑辅助设计
trace back		追溯
evolve to		进展为
incorporate	[ɪn'kɔːpəreɪt]	adj. 合并的；一体化的
composites	[kəm'pɑzɪt]	n. 复合材料（composite 的复数）
mechatronics	[mekə'trɒnɪks]	n. 机电一体化；机械电子学

nanotechnology	[ˌnænə(ʊ)tek'nɒlədʒɪ]	n. 纳米技术
aerospace engineering		［航］航空航天工程
metallurgical engineering		［冶］冶金工程
civil engineering		［建］土木工程
electrical engineering		电机工程，电气工程
manufacturing engineering		［机］制造工程
chemical engineering		化学工程
biomechanics	[ˌbaɪə(ʊ)mɪ'kænɪks]	n. 生物力学；生物机械学
subdisciplines	[sʌb'dɪsɪplɪn]	n. 学科的分支；副学科
acceleration	[əkselə'reɪʃ(ə)n]	n. 加速，促进
deformation	[ˌdiːfɔː'meɪʃ(ə)n]	n. 金属等（在压力作用下）的变形，塑性变形，拉压变形
in the most general sense		在最一般的意义上
vehicle	['viːəkl]	n. 车辆
dynamics	[daɪ'næmɪks]	n. 动力学，力学
a ventilation system		通风系统
interdisciplinary	[ˌɪntə'dɪsəplɪnerɪ]	adj. 不同学科之间的，两门以上学科间的；涉及若干（或多）学科的；跨学科的
hybrid	['haɪbrɪd]	adj. 混合的
Finite Element Analysis (FEA)		［数］有限元分析
Finite Element Method (FEM)		［数］有限元法；［数］有限单元法
viable	['vaɪəbl]	adj. 可行的

Exercise 1: Special terms.

1. aerospace engineering _____
2. metallurgical engineering _____
3. civil engineering _____
4. 电机工程，电气工程 _____
5. 制造工程 _____
6. 化学工程 _____

Exercise 2: Answer the following questions.

1. What is mechanical engineering?
2. What does the engineering field require?
3. When did mechanical engineering emerge?
4. What are subdisciplines of mechanics?

Exercise 3: Define the following terms with information from Text A.

1. mechanical engineering
2. mechatronics
3. robotics

囯 Part Two: Situational Conversations

Jordan: Hi, Can you do me a favor, Engineer Li?

Li Can: At your service.

Jordan: I don't know the exact position of the switch board. Will you scratch it on the wall?

Li Can: It should be scratched in accordance with your drawing. Where is the drawing?

Jordan: Here it is. Engineer Li.

Li Can: Sorry. You have got a wrong drawing.

Jordan: What is the problem?

Li Can: It is a plan, but what I need is an elevation, where the switch board is clearly marked. Would you please give it to me now?

Jordan: OK. Here you are.

Li Can: Thank you. (After a while) Look, the switch board is near the corner of the walls.

Jordan: You are very observant. I do admire you very much.

Li Can: If you were in my position, you'd have done the same, I am sure.

Jordan: That's true. But how high is the switch board to the floor?

Li Can: It is 1.5 m.

Jordan: How do you know the exact size?

Li Can: Because what you gave me for the second time is a dimensioned drawing. If you don't believe it, you can get a drawing scale to measure it yourself.

Jordan: In fact, there is no need at all to do that. So reading and studying carefully is an essential prerequisite for construction well, isn't it?

Li Can: Yes, absolutely. As every drawing is meticulously designed, the size and figures on the drawing are also calculated accurately what we constructors should regard is the norm and the guide of our construction works.

Jordan: That means construction is the translation of design into reality, am I right?

Li Can: Quite right. It seems that you don't read the drawing well.

Jordan: Oh, just so-so.

Li Can: Personally, I think you should make effort to read drawing well. Only in this way, can you do your work well. Don't you think so?

Jordan: Sure. I will follow your advice and redouble my efforts to make up for the lack of knowledge of building drawings.

Li Can: If you have any drawing problem in your future work and study, please don't hesitate to call me, and I will do my best to help you.

Jordan: You are very kind. Thank you very much.

Li Can: Not at all. I am honored to do something useful for you.

Exercise 1: Sentence patterns.

1. Will you do me a favor to lift the heavy box?

2. What is the problem/What is wrong/What is the matter with the machine?

3. If you were in my position, you'd have done the same, I am sure.

4. Personally, I think you'd done the same, I am sure.

5. Only in this way, can you do your work well.

6. I will follow your advice and redouble my efforts to make up for the lack of knowledge of drawing.

7. If you have any questions, don't hesitate to ask me.

8. I am honored to do something useful for you.

Exercise 2: Complete the following dialogue in English.

A: Excuse me! Can you answer me some questions about building drawings?

B: _____?

（当然可以。关于图纸你想了解什么？）

A: First, I want to know a more scientific definition of drawings.

B: _____.

（所谓图纸就是将内外形状和大小以及各部分的结构、构造、装饰、设备等内容，按照有关规范，用正投影方法详细准确地画出的图样。）

A: So drawings are very important to the construction works, aren't they?

B: _____.

（你说的没错。图纸一直被看作施工的指南和准则。）

A: How many parts are there in a complete set of mechanical drawings?

B: _____.

（一套完整的机械制图，根据专业内容或者作用不同，一般包括图纸目录、设计总说明、建筑施工图、结构施工图以及设备施工图。）

A: I know. They are mainly used in manual drawings, aren't they?

B: _____.

（没错，现在的机械制图大多是通过计算机 AutoCAD 软件绘制出来的。相对于手工绘图，其优点是不言而喻的。）

A: What is AutoCAD? I want to know.

B: _____.

（AutoCAD 是美国 Autodesk 公司开发的用于二维及三维设计、绘图的自动计算机辅助软件。）

A: Would you mind teaching me to make some drawings by AutoCAD something when you are free?

B: _____.

（当然不介意，如有需要，只管给我打电话，我一定会不遗余力地帮你。）

Part Three: Reading Comprehension

Questions 51 to 55 are based on the following passage.

A new study shows that students learn much better through an active, iterative（反复的）process that involves working through their misconceptions with fellow students and getting immediate feedback from the instructor.

The research was conducted by a team at the University of British Columbia (UBC), Vancouver, in Canada, led by physics Nobelist Carl Wieman. In this study, Wieman trained a postdoc, Louis Deslauriers, and a graduate student, Ellen Schelew, in an educational approach, called "deliberate practice," that asks students to think like scientists and puzzle out problems during class. For 1 week, Deslauriers and Schelew took over one section of an introductory physics course for engineering majors, which met three times for 1 hour. A tenured physics professor continued to teach another large section using the standard lecture format. The results were dramatic: After the intervention, the students in the deliberate practice section did more than twice as well on a 12-question multiple-choice text of the material as did those in the control section. They were also more engaged and a post-study survey found that nearly all said they would have liked the entire 15-week course to have been taught in the more interactive manner.

"It's almost certainly the case that lectures have been ineffective for centuries. But now we've figured out a better way to teach" that makes students an active participant in the process, Wieman says. The "deliberate practice" method begins with the instructor giving students a multiple-choice question on a particular concept, which the students discuss in small groups before answering electronically. Their answers reveal their grasp of the topic, which the instructor deals with in a short class discussion before repeating the process with the concept.

While previous studies have shown that this student-centered method can be more effective than teacher-led instruction, Wieman says this study attempts to provide "a particularly clean comparison…to measure exactly what can be learned inside the classroom." He hopes the study persuades faculty members to stop delivering traditional lecture and "switch over" to a more interactive approach. More than 55 courses at Colorado across several departments now offer that approach, he says, and the same thing is happening gradually at UBC.

51. What do we know about the study led by Carl Wieman in the second paragraph?

   A) Students need to turn to scientists for help if they have trouble.

   B) An introductory physics course was given to physics majors.

   C) Students were first taught in the " deliberate practice "approach.

   D) A professor continued to teach the same section with the traditional lectures.

52. The results of the research reveal that _____.

   A) the students in the experimental section performed better on a test

   B) the students in the control section seemed to be more engaged

   C) the students preferred the traditional lectures to deliberate practice

   D) the entre 15-week course was actually given in the new manner

53. How does Wieman look at the traditional lectures according to the third paragraph?

   A) They have lasted for only a short period of time.

   B) They continue to play an essential role in teaching.

   C) They can make students more active in study.

   D) They have proved to be ineffective and outdated.

54. How does the "deliberate practice" method work?

   A) The students are first presented with some open questions.

   B) The students have to hand in paper-based homework.

   C) The instructor remains consistent in the way of explaining concept.

   D) The instructor expects the students to air their views at any time.

55. We learn from the last paragraph that Wieman's new approach _____.

   A) can achieve the same effects as the traditional lecture format

   B) can evaluate the students' class performance roughly

   C) will take the place of the traditional way of teaching in time

   D) has been accepted by faculty members in some colleges

Part Four: Translation Skills

翻译技巧：省译法

省译法是指在翻译中舍去原文中那些可有可无的成分。省译目的在于使译文更加流畅，符合目的语的习惯。英译汉时省略的大多是英语中因为语法上的需要而存在，但根据汉语习惯并不需要译出来的词语。因此，省译的原则是：省译的部分在译文中不言而喻，译出来反而拖沓累赘；省译后不能有损原意或改变原文色彩。

具体而言，省译法包括省略原文中的词语、省略语法范畴以及简化累赘的文体。这是与增译法相对应的一种翻译方法，即删去不符合目标语思维习惯、语言习惯和表达方式的词，以避免译文累赘。如：

例1： He puts his hand into his pockets and then shrugged his shoulders.

翻译： 他把双手放进口袋里，然后耸耸肩。（省译代词）

详解： 本例句如果直译出来是"他把他的双手放进了他的口袋里，然后耸耸他的肩"。这样翻译就显得语句有些拖沓、累赘，因此可以省译三个物主代词"his"，译为"他把双手放进口袋里，然后耸耸肩。"这样意思明确，读起来也较通顺、自然，符合汉语习惯。

例2： The report began, and the audience stopped talking.

翻译： 报告开始了，听众停止了谈话。（省译连词）

详解： 英语中语句的逻辑关系往往是通过连词明显地表现出来，然而中文语句间的关系往往是内隐的。因此英译汉时，就需要适当省译连词。如例2中的"and"，这种顺接关系在汉语中是不需要明确表现出来的，故省译该连词。

例3： He hit her in the face.

翻译： 他打了她的脸。（省译介词）

详解： 有时根据汉语习惯并不需要译出介词，比如例3如果直译就是"他打进了她的脸"，这样读者就会不知所云。所以此处需要省译介词"in"，译作"他打了她的脸"。

Translation Exercise 1

机械工程在18世纪欧洲的工业革命中应运而生，然而，它的发展可以追溯到几千年前的世界各地。机械工程在19世纪的物理学领域出现进而发展，随后在这一领域里技术得到不断进化和进步，今天的机械工程师们追求复合材料、机电一体化和纳米技术等方面的发展。机械工程与航空航天工程、冶金工程、土木工程、电气工程、制造工程、化学工程等其他工程学科不同，机械工程师也可能在生物医学工程领域工作，尤其是在生物力学、交通现象、生物和生物系统的建模方面。

Translation Exercise 2

排除故障的步骤：在未确定具体故障的情况下，可以先查看连接端口的 E1 线缆是否状况良好，连接好后查看报警是否消除。如果报警还存在，可以通过对 E1 盘进行设备环回和线路环回测试确定出现故障的机盘。

Unit 3 Industrial Robots

Learning Objectives

After completing this unit, you will be able to do the following:

1. Grasp the main idea and the structure of the text;
2. Master the key language points and grammatical structure in the text;
3. Understand the basic concepts in industrial robots;
4. Conduct a series of reading, listening, speaking and writing activities related to the theme of the unit.

Technical Terms

In this unit, you will learn some technical terms in industrial robotics listed below:

- robotics;
- industrial robots;
- gripper;
- jaw;
- built-in control system;
- endpoint of the cutter;
- programmable mechanical manipulator.

Outline

The followings are the main sections of this unit:

1. Warm-up Activity;
2. Text A ;
3. Situational Conversation;
4. Reading Comprehension;
5. Translation Skills;
6. Excercises.

Vocabulary

The listed below are some words appearing in this unit that you should make part of your vocabulary:

- algorithm;
- consequently;
- velocity;
- weld;
- wrist;
- axe;
- gripper;
- rinse.

🖝 Looking Ahead

Robotics

The word robotics was derived from the word robot, which was introduced to the public by Czech writer Karel Čapek in his play *R. U. R. (Rossum's Universal Robots)*, which was published in 1920. The word *robot* comes from the Slavic word *robota*, which means labour. The play begins in a factory that makes artificial people called *robots*, the creatures which can be mistaken as humans—very similar to the modern ideas of androids. Karel Čapek himself did not coin the word. He wrote a short letter in reference to an etymology in the *Oxford English Dictionary* in which he named his brother Josef Čapek as its actual originator.

According to the *Oxford English Dictionary*, the word *robotics* was first used in print by Isaac Asimov, in his science fiction short story *Liar*, published in May 1941 in *Astounding Science Fiction*. Asimov was unaware that he was coining the term; since the science and technology of electrical devices is *electronics*, he assumed that robotics had already been referred to the science and technology of robots. In some of Asimov's other works, he states that the first use of the word *robotics* was in his short story *Runaround* (*Astounding Science Fiction*, March 1942). However, the original publication of *Liar* predates that of *Runaround* by ten months, so the former is generally cited as the word's origin.

☕ Core Contents

1. The discussion of the definition of robot.
2. Discussion of the operation and application of robot in the manufacturing.

◆ Warm-up Activity

What's your impression of a robot?

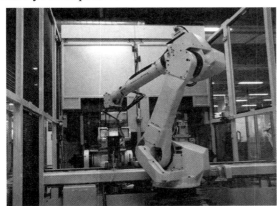 

Manipulator   Intelligent robot

 Part One: Text A Basic Concepts in Robotics

Industrial robots begin to revolutionize industry. These robots do not look or behave like human beings, but they do the work of humans. Robots are particularly useful in a wide variety of application, such as material handling, spray painting, spot welding, arch welding, inspection, and assembly. Current research efforts focus on creating a "smart" robot that can "see", "hear" and "touch" and consequently make decisions.

The technology of robots is not only related to, but different somehow from, NC technology in that robots effect higher velocity and movement in more axes of motions. While with NC only a point, namely the endpoint of the cutter, is controlled in the space, with robots both the endpoint and the orientation are manipulated. This requires more degrees of freedom, more powerful software, and more effective control algorithms.

The industrial robot is a programmable mechanical manipulator, capable of moving along several directions, equipped at its end with a device called the end-effector, and performs factory work ordinarily done by human beings. The term robot is used for a manipulator that has a built-in control system and is capable of stand-alone operation.

Webster's Dictionary defines a robot as "any mechanical device operating automatically to perform in a seemingly human way." By this definition, a garage door opener, which automatically opens the door by remote control is also a robot. Obviously this is not an industrial robot. The Robot Institute of America (RIA) defines the industrial robot as "a reprogrammable multi-functional manipulator designed to move materials, parts, tools, or other specialized devices through variable programmed motions for the performance of a variety of tasks." By this definition, however, a washing machine is also a robot. The wash and rinse cycles are programmable and the machine moves materials in rotary motions. Therefore, a definition of an industrial robot must include the following key words: programmable manipulator, end-effector, factory work, and stand-alone operation. If one or more of these key words are missing, then washing machine, traffic lights, special purpose mechanisms and manufacturing machines for mass production are defined as robots as well, and this is not our intention.

In general, the structure of a robot manipulator is composed of a mainframe and a wrist at its end. The main frame is frequently denoted as the arm, and the most distal group of joints affecting the motion of the end-effector is referred to collectively as the wrist. The end-effector can be a welding head, a spray gun, a machining tool, or a gripper containing on-off jaws, depending upon the specific application of the robot. Each of these devices is mounted at the end of the robot and performs the work, and therefore is also denoted as the robot tool.

Words and Expressions

algorithm	['ælgərɪðəm]	n. 算法
axe	[æks]	n. 轴
consequently	['kɒnsɪkwəntlɪ]	adv. 所以，因而
denote	[dɪ'nəut]	v. 记作，称为
gripper	['grɪpə]	n. 手爪
jaw	[dʒɔ:]	n. 爪钳夹装置
rinse	[rɪns]	n. 冲洗、漂洗
velocity	[vɪ'lɒsətɪ]	n. 速度，速率
weld	[weld]	n. 焊接
wrist	[rɪst]	n. 腕，腕关节
be defined as		（被）定义为
be equipped with		安装有……，装备有……
behave like		举止、行为像……
built-in control system		内置控制系统
degrees of freedom		自由度
differ from		与……不同
endpoint of the cutter		末端切削刀具
in a seemingly human way		以类人的方式
industrial robot		工业机器人
make decisions		（做）决策
material handling		材料搬运

remote control	遥控
RIA (The Robot Institute of America)	美国机器人学会（研究所）
special purpose mechanisms	专用机械装置
spot welding	点焊
spray painting	喷涂
welding inspection	焊接质量检验
programmable mechanical manipulator	可编程机械装置
traffic lights	交通灯
welding head	焊头
spray gun	喷枪

Notes:

1. Current research efforts focus on creating a "smart" robot that can "see" "hear" and "touch" and consequently make decisions.

本句的意思是：当前研究主要集中在建造能"看"、能"听"、能"触摸"的"聪明的"机器人上。此处的"smart"（聪明的）robot 指的是我们通常所说的"智能型机器人"。

2. gripper containing on-off jaws

上面装有抓放机构的钳爪

3. If one or more of these key words are missing, then washing machines, traffic lights, special purpose mechanisms and manufacturing machines for mass production are defined as robots as well, and this is not our intention.

本句是对《韦伯斯特词典》（*Webster's Dictionary*）有关机器人的定义的反驳，意思是：如果一个或多个关键词被省略，那么洗衣机、交通灯、专用机床等都可以叫机器人，这并不是我们的本意所在。

Exercise 1: Special terms.

1. robot   _____

2. NC   _____

3. intelligent robot   _____

4. 可编程的   _____

5. 内置控制系统   _____

6. 遥控   _____

Exercise 2: Answer the following questions.

1. What's the difference between a NC and a robot?

2. According to the definition of The Robot Institute of America (RIA), what is an industrial robot?

3. What tasks can an industrial robot do when equipped with proper tools?

4. Can a washing machine be called a robot?

Exercise 3: Define the following terms with information from Text A.

1. industrial robot

2. intelligent robot

3. manipulator

📑 Part Two: Situational Conversation

Topical introduction: Tom and Mary are now visiting an industrial robot show in a university. A young girl named Marcy is in charge of showing the visitors around.

Marcy: Ladies and gentlemen, welcome to our industrial robot show. I'm Marcy Margaret. My job is to show you around. You can ask me about anything you do not understand.

Mary: How interesting to see so many kinds of robots here. Excuse me, Marcy, but I wonder why these robots do not look like or behave like human beings?

Marcy: Ah, I see. The point here is that what we show you are not other kinds of robots but industrial ones. We needn't make them look like us humans.

Mary: Well, I understand. Now could you tell us what these industrial robots can do for us?

Marcy: Sure. These robots can do many things for us. They can help us, for example, to handle materials, spray paint, rescue people from the fire, and even help to explore deep oceans and outer-space.

Tom: Uh, how great robots are! But what makes robots able to work for us so marvelously?

Marcy: It's nothing else but the computer programs. Robots are a programmable mechanical manipulator. Being programmed with human instructions, they can move along several directions and do factory work usually done by human beings.

Mary: Oh, that's wonderful. But what's the difference between robot and my handy calculator or my home washing machine, since all of them belong to some reprogrammable mechanical manipulator.

Marcy: Okay, it's a clever question. But I think, here you mix a robot with an ordinary washing machine because you forget a robot can do a stand-alone operation, that is, rather independent jobs, while a washing machine can't. So, when we talk about industrial robots, we mean a machine or device, which is something reprogrammable with its own end effector, and can perform a factory duty in a stand-alone manner.

Tom: Oh, I see. I guess, my automatic watch and my home washing machine cannot play chess with me while a robot can.

Marcy: That's right. You're a clever boy, I see.

Notes:

spray painting 喷涂

marvelously 奇迹般地，奇异地

programmable mechanical manipulator 可编程机械装置

Exercise 1: Read the following dialogue and filled the blanks with the suggested expressions.

Suggested expressions:

1. Yes, I am

2. I guess we can

3. which can walk and blink to us like a real child

4. do not look like

Mary: (Turning to Tom) Tom, are you interested in the industrial robot show we visited yesterday?

Tom: _____.

Mary: So am I. But I wonder why these industrial robots _____ or behave like us real human beings?

Tom: Perhaps they needn't be so. I think, it would be much easier for the robot experts to make us a beautiful doll _____.

Mary: Ah, I see. But can we hope to find robots in some other robots shows which do look like human beings?

Tom: Well,_____

Exercise 2: Read the following dialogue and filled the blanks with the suggested expressions.

Suggested expressions:

1. in the dictionary

2. should be used as a tool

3. to discuss with you

4. the meaning

5. would be a robot then, wouldn't be

Mary: Good morning, Tom, did you review the text?

Mary: Excuse me, Tom. Could you spare me a few minutes? I have something on robot and robotics _____.

Tom: Sure. Why not?

Mary: Yesterday when I was reviewing the text on the industrial robots, I was trying to look up _____ of robot _____. I found something confusing me.

Tom: What does the dictionary say about robot?

Mary: It defines a robot as "anything mechanical device operated automatically to perform in seemingly human way." If so, a hotel automatic gate _____?

Tom: No, never. So I don't think we should memorize all the definitions and explanations in it even if it is a good dictionary. Rather, a dictionary _____.

▶ Part Three: Reading Comprehension

That orientals and westerners think in different ways is not mere prejudice. Many psychological students who conducted over the past two decades suggest that westerners have a more individualistic and abstract mental life than East Asians do. Several expectations are proposed to account for different ways of thinking.

One explanation is that stepping into the modern social, economic and technological situation promotes individualism. However, in Japan, it is pretty liable to a higher frequency of infectious disease; it is more dangerous to make contact with strangers, which causes groups in this place to turn inward and tend to be collective. This explanation is also questioned. Europe has had its share of plagues, probably more than either Japan or Korea.

That led Thomas Talhelm of the University of Virginia and his colleagues to look into a third suggestion: that the crucial difference is agricultural. The west's staple is wheat, while the east's is rice. Before that the crucial difference is agricultural, a farmer who grew rice had to spend twice as many hours as one who grew wheat to promote efficient agricultural production, especially at times of planting and harvesting, rice-growing societies as far apart as India, Malaysia and Japan all developed cooperative labor exchanges. That is, neighbors arranged their farm schedules one after another in order to assist each other during these significant periods. Since, until recently, almost everyone is a farmer, it is a reasonable proposal that such a collective outlook would enjoy a controlling position in society's culture and behavior, and might prove to be deep rooted even now, and when most people earn their living in other ways, it helps to define their lives. This proposal that the different ways of thinking of the East and West are, at least in part, a consequence of their agriculture is worth future exploration.

1. What do we learn about easterners' way of thinking?

   A) They often hold no prejudice.

   B) They pursue abstract mental life.

   C) They are in favor of individualism.

   D) They emphasize the collective values.

2. What does the first explanation indicate?

   A) Individualism has little relationship with the modern development.

   B) Individualism is inspired by modern society, economy and technology.

   C) Japan can be counted as a modern country.

   D) The Japanese prefer a more collective life.

3. Which of the following questions is the second proposal?

   A) People in a place with infectious disease tend to advocate individualism.

   B) People in a place with infectious disease tend to contact with outsiders.

   C) In China, it is easy to prevent the spread of infectious disease.

   D) Europe is more likely to be infected by plagues than Korea.

4. What do Thomas and his colleagues assume?

   A) The cultivation of wheat requires twice as much time as that of rice.

   B) In Malaysia, farmers need their neighbors' assistance to grow wheat.

   C) The major agricultural crop of rice may define cooperative outlook.

   D) Farmers are only engaged in their own planting and harvesting of rice.

5. The author's attitude toward Thomas and his colleagues' suggestion is _____.

   A) indifferent   B) objective

   C) critical   D) disappointed

Part Four: Translation Skills

翻译技巧：重复法

重复法（repetition）是在翻译中为了使译文忠实于原文并且产生意义明确，文字通顺、流畅，符合目的语习惯的文字，而将某一部分文字反复使用的翻译技巧。在译文中适当地重复原文中出现过的词语，以使意思表达得更加清楚；或者进一步加强语气，突出强调某些内容，都会收到更好的修辞效果。一般情况下，英语常常会为了行文简洁而尽量避免重复。因此，经常借助替代、省略或变换等其他表达方法。与此相反，重复是汉语表达的一个显著特点。在许多场合某些词语不仅需要重复，而且也只有重复这些词语，语义才能明确，表达才能生动。为了达到汉语译文准确、通顺和完整的翻译标准，在英译汉中，经常会采用重复法，如：

例 1： You must ask the mother at home, the children in the street, the ordinary man in the market and look at their mouth, how they speak, and translate that way; then they'll understand and see that you're speaking to them in German.

翻译： 你一定要问一问家庭主妇们，问一问街头玩耍的孩子，问一问集市上做买卖的百姓，听听他们说些什么，他们如何说，你就如何译；这样他们就会理解，就会明白：你是在用德语和他们讲话。（重复谓语动词）

详解： 该句的中文翻译将原文的谓语动词 ask 进行了三次重复，由于 ask 的宾语各不相同，如果不将谓语部分重复表达，会影响"问一问"宾语对象的明确性。

例 2： We have to analyze and solve problems.

翻译： 我们必须分析问题，解决问题。（重复宾语）

详解： 在本句的翻译中，将句子中的宾语"problems"进行了重复翻译，其目在于能够将 analyze 和 solve 的宾语交代清楚，如果不加以重复，可能会对分析的对象产生疑问。

例 3： Ignorance is the mother of fear as well as of admiration.

翻译： 无知是畏惧之源，羡慕之根。（重复省略的部分）

详解： 本句的英文原文中有一个省略成分，也就是省略了 as well as 后面的 mother，在英语中，这种省略方式可以避免表达上的重复和拖沓。而在中文中如果采取同样的方式，就会译为"无知是畏惧也是羡慕之根"，会产生一定的歧义。因此，为避免不必要的歧义，应对省略部分进行重复翻译。

Translation Exercise 1

工业机器人正在使工业发生革命性的变化。它们从外貌上不像人，也表现得不像人，但它们确实做着人类的工作。机器人在很多应用领域里的用处特别明显，如材料处理、喷涂、点焊、焊接、检查和装配等。目前的研究工作重点集中在能创造一个"智能"机器人，它能"看"，能"听"，也能"触摸"，从而作出决定。

Translation Exercise 2

保修期内，用户送修时必须持有购买送修产品的有效发票和厂方指定的相关三包凭证，"三包"有效期自发票开具之日算起。用户应妥善保存保修卡和购机发票，送修时必须同时出示保修卡和购机发票。用户遗失购买发票时，应按出厂日期推算"三包"有效期。

Unit 4 Automation in Manufacturing

Learning Objectives

After completing this unit, you will be able to do the following:

1. Grasp the main idea and the structure of the text;
2. Master the key language points and grammatical structure in the text;
3. Understand the differences between automation and mechanization;
4. Conduct a series of reading, listening, speaking and writing activities related to the theme of the unit.

Outline

The followings are the main sections of this unit.

1. Warm-up Activity;
2. Text A;
3. Situational Conversation;
4. Reading Comprehension;
5. Translation Skills;
6. Exercises.

Technical Terms

In this unit, you will learn some technical terms about automation in manufacturing listed below:

- automation;
- mechanization;
- feedback;
- pneumatic;
- hydraulic;
- centrifugal;
- shaft;

Vocabulary

The listed below are some words appearing in this unit that you should make part of your vocabulary:

- discrimination;
- feedback;
- deviation;
- remarkable;
- prestige;
- exceptional;
- immense;
- innumerable.

Looking Ahead

Automation

The term *automation*, inspired by the earlier word *automatic* (coming from *automation*), was not widely used until 1947, when General Motors established an automation department. It was during this time that industry rapidly adopted feedback controllers, which were introduced in the 1930s.

Automation has been achieved by various means including mechanical, hydraulic, pneumatic, electrical, electronic devices and computers, usually in combination. Complicated systems, such as modern factories, airplanes and ships typically use all these combined techniques.

Core Contents

1. The differences between automation and mechanization.
2. The essence of an automatic mechanism.

◆ Warm−up Activity

Have you ever visited an high-technology company? Share your experience with your classmates.

Robot in manufacturing

Automation in manufacturing

▣ Part One: Text A The Differences Between Mech−anization and Automation

Processes of mechanization have been developing and becoming more complex ever since the beginning of the Industrial Revolution at the end of the 18th century. The current development of automatic processes is, however, different from that of the old ones. The "automation" of the 20th century is distinct from the mechanization of the 18th and 19th centuries in as much as mechanization was applied to individual operations, whereas "automation" is concerned with the operation and control of a complete producing unit. And in many, though not all, instances the element of control is so great that mechanization displaces muscle, and "automation" displaces brain as well.

The distinction between the mechanization of the past and now is, however, not a sharp one. At one extreme we have the electronic computer with its quite remarkable capacity for discrimination and control, while at the other end of the scale is "transfer machines", as they are now called, which may be as simple as a conveyor belt to another. An automatic mechanism is one which has a capacity for self-regulate; that is, it can regulate or control the system or process without the need for constant human attention or adjustment.

Now people often talk about "feedback" as being an essential factor of the new industrial techniques, based upon which is an automatic self-regulating system and by virtue of which any deviation in the system from desired conditions can be detected, measured, reported and corrected. When "feedback" is applied to the process by which a large digital computer runs at the immense speed through a long series of sums, constantly rejecting the answers until it finds one to fit a complex set of facts that have been put to it, it is perhaps different in degree from the machines that we have previously been accustomed to. But "feedback", as such, is a familiar mechanical conception. The old-fashioned steam engine was fitted with a centrifugal governor, and two balls on levers spinning round and round an upright shaft. If the steam pressure rose and the engine started to go too fast, the increased speed of the spinning governor caused it to rise up the vertical rod and shut down a valve. This cut off some of the steam and thus the engine brought itself back to its proper speed.

The mechanization, which was introduced with the Industrial Revolution, because it was limited to individual processes, required the employment of human labor to control each machine as well as to load and unload materials and transfer them from one place to another. Only in a few instances were processes automatically linked together and was production organized as a continuous flow.

In general, however, although modern industry has been highly mechanized ever since the 1920s, the mechanized parts have not been linked together as a rule. Electric-light bulbs, bottles and the components of innumerable mass-produced articles are made in mechanized factories in which a degree of automatic control has gradually been building up. The development of the electronic computers in the 1940s suggested that there were a number of other devices less complicated and expensive than the computer which could share the field of mechanical control. These mechanical, pneumatic and hydraulic devices have been considerably developed in recent years and will continue to advance now that the common opinion is favoring the extension of "automation". Although electronic devices, of course, are not the sole because of what is happening, they are nevertheless in a key position. They are gaining in importance and unquestionably hold out exceptional promise for development in the future.

Words and Expressions

mechanization	[ˌmekənaɪˈzeiʃən]	n. 机械化
automation	[ɔ:təˈmeɪʃən]	n. 自动化
in as much as		因为，由于，鉴于，既然
discrimination	[dɪsˌkrɪmɪˈneɪʃən]	n. 辨别，鉴别，识别，歧视
feedback	[ˈfi:dbæk]	n. 反馈，回馈
negative feedback		负反馈

deviation	[dɪˌviˈeiʃən]	*n.* 偏差，偏移，背离
spinning	[ˈspɪnɪŋ]	*adj.* 旋转的 *n.* 纺线，纺纱
pneumatic	[njuːˈmætɪk]	*adj.* 空气的，气动的
pneumatic cylinder		气缸
pneumatic actuator		气动执行元件
hydraulic	[haɪˈdrɔlɪk]	*adj.* 液压的
hydraulic engineering		水利工程
hydraulic drive		液压传动
hydraulic transmission system		液压传动系统
remarkable	[rɪˈmɑːkəbl]	*adj.* 显著的，非凡的，值得注意的
remarkable achievements		卓越的成就
prestige	[preˈstiːʒ]	*n.* 名望，声望，威望
displace	[dɪsˈpleɪs]	*vt.* 移置，取代，转移　过去式：displaced 过去分词：displaced　现在分词：displacing
exceptional	[ɪkˈsepʃənl]	*adj.* 例外的，异常的，特殊的，优秀的，卓越的
old-fashioned	[ˈəuldˈfæʃənd]	*adj.* 老式的，旧的
immense	[ɪˈmens]	*adj.* 极大的，巨大的
centrifugal	[senˈtrɪfjuɡəl]	*adj.* 离心的
centrifugal force		离心力
centrifugal governor		离心式调速器
shaft	[ʃɑːft]	*n.* [C] 轴；矿井，竖井
valve	[vælv]	*n.* 阀，活门；真空管，电子管
innumerable	[ɪˈnjuːmərəbl]	*adj.* 无数的；数不清的
mass-produced		大（批）量生产的

Exercise 1: Special terms

1. mechanization　_____

2. automation　_____

3. mass-produced articles　_____

4. 自我调节　_____

5. 液压的　_____

6. 装运和卸载　_____

Exercise 2: Answer the following questions.

1. Is the "automation" of the 20th century same with the mechanization of the 18th

and 19th centuries?

    2. What can an automatic mechanism do in the actual manufacturing?

    3. What does "feedback" mean in the automation in manufacturing?

    4. When has the modern industry been highly mechanized?

Exercise 3: Define the following terms with information from Text A.

    1. automation

    2. transfer machines

    3. mechanization of the Industrial Revolution

Part Two: Situational Conversation

Tom: Hello, Professor Jackson. We've been admiring you for years. We are so glad to meet you here in this seminar.

Jackson: Oh, yes, young man, glad to meet you too.

Tom: Oh, Professor. Would you mind if I ask you some questions on your topic today? Automation and Robotics?

Jackson: Of course. Come on then.

Tom: Thank you. My first question is something on the general idea with robots. In my mind, robots are no more than a manufacturing tool. They are used to do things we human beings do not like to do, while automation is by nature a production means involving the production organization. So I can't see the point why you call them two closely related technologies.

Jackson: Oh, I see. What you said sounds reasonable to some extent. But both of them would be very common if we take them as tools or means that help to reduce the workload of us human beings.

Tom: Yes, I understand. One more thing I want to ask you. In the seminar, some people mentioned the risk or danger that would occur with fixed automation. Could you explain to us why there would be risk involved with the kind of automation?

Mary: Oh, yes. I have the same question, too.

Jackson: Good. The risk you mentioned actually involves long-time production planning strategy. You know, it is very expensive to set up highly integrated automatic transfer lines, automatic processing or matching and assembly lines. Once all such lines are built up, you must find plenty of work for them to do. If you can't, there would be an enormous waste. After all, such production lines cannot be changed or moved easily. Can't you see?

Tom and Mary: Oh, yes. That is what fixed automation means.

Jackson: Exactly.

Mary: Professor Jackson, I was puzzled when people in the discussion said that of the three types of automation, robots coincide most closely with programmable automation. Could you tell us why?

Jackson: OK, no problem. Needless to say, robots are useful in all kinds of automation. For example, in automobile plants, robots are used to move materials or work on transmission components. But, by nature, a robot is nothing but a programmed tool or device or technology just like the programmable automation. Both of them are working at the programs laid out by the experts of artificial intelligence.

Tom and Mary: Oh, yes. We see what you mean.

Notes:

1. seminar [ˈsemɪnɑː(r)] *n.* 研讨班，讲习会；研讨小组；研讨会；培训会
2. robotics [rəʊˈbɒtɪks] *n.* 机器人学
3. no more than 不过；仅仅

Exercise 1: Read the following dialogue and filled the blanks with the suggested expressions.

Suggested expressions:

1. why not

2. parts and components

3. the line

4. varieties

Mary: (Tom and Mary now in an assembly shop of an auto-plant) Tom, look how fast the assembly work is done on the line.

Tom: Yes, marvelous. Oh, Mary, let's go over to the shop guy and ask something about _____. OK?

Mary:_____? (Mary is now turning to a shopworker) Good morning, master! Can I ask you some questions?

Shopworker: Yes, you are welcome.

Tom: May I ask you how many motorcycles can you put together everyday?

Shopworker: About 300.

Mary: Wonderful. If 300 in a day, you can assemble 100,000 _____ in a year.

Shopworker: Theoretically, yes. But we have often to sit and wait for more jobs to do. Because there are not so many _____ to assemble.

Tom: If so, can you change your assembly line somewhat for the new uses?

Shopworker: I'm sorry. It's almost impossible. You know, this assembly line is fixed.

Exercise 2: Role-play the above conversation and add some necessary information if possible.

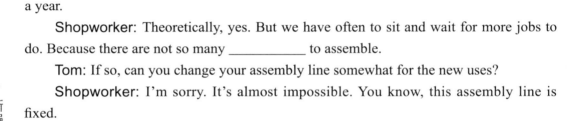

Part Three: Reading Comprehension

When students arrive on campus with their parents, both parties often assume that the school will function in loco parentis（处在父母位置）, watching over its young charges, providing assistance when needed. Colleges and universities present themselves as supportive learning communities—as extend families, in a way. And indeed, for many students they become home away from their real home. This is why graduates often use another Latin term, alma mater, meaning "nourishing mother". Ideally, the school nurtures its students, guiding them toward adulthood. Lifelong friendship is formed; teachers become mentors（导师）; and the academic experience is complemented by rich social interaction. For some students, however, the picture is less rosy. For a significant number, the challenges can become overwhelming.

In reality, administrators at American colleges and universities are often obliged to focus as much on the generation of revenue as on the new generation of students. A trouble or even severely disturbed student can easily fall through the cracks. Public institutions in particular are often faced with tough choices about which student support services to fund, and how to manage such things as soaring health-care cost for faculty and staff. Private schools are feeling the pinch as well. Ironically, although tuition and fees can increase as much as 6.6 percent in a single year, as they did in 2007, the high cost of doing business at public and private institutions means that students are not necessarily receiving more support in return for increased tuition and fees. To compound the problem, students may be reluctant to seek help even when they desperately need it.

Unfortunately, higher education is sometimes more of an information delivery system than a responsive, collaborative process. Just as colleges are sometimes ill equipped to respond to the challenges being posed by today's students, so students themselves are sometimes ill equipped to respond to the challenges posed by college life. Although they arrive on campus with high expectations, some students struggle with chronic shyness, learning disabilities, addiction, or eating disorders, while others suffer from acute loneliness, mental illness, or even rage.

We have created cities of youth in which students can pass through unnoticed; their voices rarely heard; their faces rarely seen. As class size grows in response to budget cuts, it becomes even less likely that troubled students, or even severely disturbed students, will be noticed. When they're not, the results can be tragic.

1. How do students feel about colleges and universities ?

    A) Admiring.   B) Disappointed.   C) Indifferent.   D) Affectionate.

2. What's the ideal image of colleges and universities ?

    A) They are places where academic requirements are loose.

    B) They are places where students can have colorful social experience.

    C) They nurture students and guide them to grow into adults.

    D) They teach students how to spend their youth time best.

3. Why do American colleges and universities often neglect troubled or even severely disturbed students ?

    A) They view academic management as their only task.

    B) They don't have so much energy or money to focus on the needs of students.

    C) They don't care about the students' mental health.

    D) They focus mainly on improving the salaries for faculty and staff.

4. Which of the following information can be got from the third paragraph ?

    A) Private schools in American never feel that they are short of money.

    B) American students receive more support as tuition and fees increase.

    C) American colleges and universities fail to respond to and help students in time sometimes.

    D) American college students seek help only when they suffer from serious mental illness.

5. What's the author's attitude towards American higher education?

    A) Critical.   B) Neutral.   C) Praising.   D) Unintended.

Part Four: Translation Skills

翻译技巧：转换法

转换法指翻译过程中为了使译文符合目标语的表述方式、方法和习惯而对原句中的词类、句型和语态等进行转换。具体来说，就是在词性方面，把名词转换为代词、形容词、动词；把动词转换成名词、形容词、副词、介词；把形容词转换成副词和短语。在句子成分方面，把主语转换成状语、定语、宾语、表语；把谓语转换成主语、定语、表语；把定语转换成状语、主语；把宾语转换成主语。在句型方面，把并列句转换成复合句，把复合句转换成并列句，把状语从句转换成定语从句。在语态方面，可以把主动语态转换为被动语态。如：

例1： The reform and opening policy is supported by the whole Chinese people.

翻译：改革开放政策受到了全中国人民的拥护。（动词转名词）

详解：在本句的翻译中，原文中的谓语动词结构 is supported by 在翻译中被转换成了名词结构"拥护"。这样的翻译转换处理是由于在汉语中不太习惯用被动意义来表达对于政府的支持，而"受到全国人民的拥护"更符合汉语的表达习惯，更有利于读者的理解。

例2： In his article the author is critical of man's negligence toward his environment.

翻译：作者在文章中，对人类疏忽自身环境作了批评。（形容词转名词）

详解：本句的翻译中，原文中的形容词结构 critical of 被转化为了名词结构"批评"。之所以这样处理，是因为在汉语中，对于批评这个意义主要限定于动词和名词词性，因此，将这个形容词结构转化为名词结构，更符合汉语表达习惯。

例3： In some of the European countries, the people are given the biggest social benefits such as medical insurance.

翻译：在有些欧洲国家里，人民享受最广泛的社会福利，如医疗保险等。（被动语态转主动语态）

详解：在中文里，"给予"这个词通常用作主动含义，而被动意义很少使用，也不太符合中文的表达习惯，另外该句英文中"被给予最大的社会福利"从意义上来说，表达的就是"人民能享受到最广泛的社会福利"。因此在翻译时，将 are given the biggest social benefits 转换为了主动语态。

Translation Exercise 1

自从 18 世纪末工业革命开始，工业机械化进程一直在不断地发展并且变得越来越复杂。但目前的工业自动化过程与以前的工业自动化过程相比有很大的不同。20 世纪的工业自动化之所以有别于 18 世纪和 19 世纪的机械化，是因为机械化仅应用于操纵（执行）机构，而自动化则涉及整个生产单元的执行和控制两个（核心）部分。在大多数情况下，控制元件依然发挥着强大的作用，机械化已经代替了手工劳动，而自动化也代替了脑力劳动。

Translation Exercise 2

高清晰度：采用 MPEG2 编码格式，使水平清晰度达到 500 线以上。

时间搜索：可快速找到碟片上某一点的内容，尤其适合武打故事片的欣赏。

内容显示：采用彩色荧光显示及电视屏中英文显示，碟片信息一目了然。

Unit 5 Mechatronics

Learning Objectives

After completing this unit,you will be able to do the following:

1. Grasp the main idea and the structure of the text;
2. Master the key language points and grammatical structure in the text;
3. Understand the essence of mechatronics;
4. Conduct a series of reading, listening, speaking and writing activities related to the theme of the unit.

Outline

The followings are the main sections of this unit.

1. Warm-up Activity;
2. Text A;
3. Situational Conversation;
4. Reading Comprehension;
5. Translation Skills;
6. Excercises.

Technical Terms

In this unit, you will learn some aviation terms listed below:

* mechatronics;
* mechanics;
* electronics;
* modelling;
* sensor;
* optoelectronic;
* actuator;
* synergistic.

Vocabulary

The listed below are some words appearing in this unit that you should make part of your vocabulary:

* interdisciplinary;
* automotive;
* intelligent;
* forward-thinking;
* encompass;
* transaction;
* integration;
* category;
* fusion;
* adaptable;
* coin;
* array.

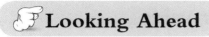

Looking Ahead

Mechatronics

The word "mechatronics" originated in Japanese-English and was created by Tetsuro Mori, an engineer of Yaskawa Electric Corporation. The word "mechatronics" was registered as trademark by the company in Japan with the registration number of "46-32714" in 1971. However, afterwards the company released the right of using the word to public, and the word "mechatronics" spread to the rest of the world. Nowadays, it is translated in every language and is considered as an essential term for industry.

Core Content

1. The development and interdisciplinary characteristics of the subject;
2. The essence of mechatronics.

Warm-up Activity

Being labeled as a "world factory" for so many years, China lays much emphasis on the subject, Mechatronics. Why?

Digital camera

Washing machine

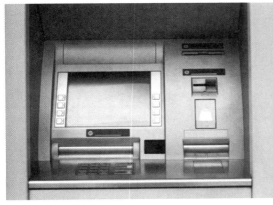

ATM

Printer

 Part One: Text A Mechatronics

Mechatronics is nothing new; it is simply the application of the latest techniques in precision mechanical engineering, control theory, computer science, and electronics to the design process to create more functional and adaptable products. This, of course, is something that many forward-thinking designers and engineers have been working on for years.

As shown in the figure below, mechatronics is the interdisciplinary fusion (not just a simple mixture!) of mechanics, electronics and information technology. The objective is for engineers to complete development, which is why it is currently so popular with industry.

A Japanese engineer from Yasukawa Electric Company coined the term "mechatronics" in 1969 to reflect the merging of mechanical and electrical engineering disciplines. Until the early 1980s, mechatronics meant a mechanism that was electrified. In the mid-1980s, mechatronics came to mean engineering which was the boundary between mechanics and electronics.

Today, the term encompasses a large array of technologies, many of which have become well-known in their own fields. Each technology still has the basic element of the merging of mechanics and electronics but now many also involve much more, particularly software and information technology. For example, many early robots resulted from mechanical and electrical systems became central to mechatronics.

Mechatronics gained legitimacy in academia in 1996 with the publication of the first referred journal: *IEEE/ASME Transactions on Mechatronics*. In the premier issue, the authors worked to define mechatronics. After acknowledging that many definitions have circulated, they selected the following for articles to be included in it: "The synergistic integration of mechanical engineering with electronics and intelligent computer control in the design and manufacture of industrial products and processes."

The authors suggested 11 topics that should fall, at least in part, under the general category of mechatronics:

Modelling and design;
System integration;
Actuators and sensors;
Intelligent control;
Robotics;
Manufacturing;
Motion control;
Vibration and noise control;
Microelectronic devices and
optoelectronics systems;
Automotive systems; and
Other applications.

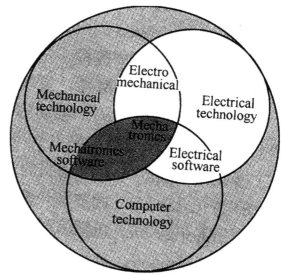

The interdisciplinary nature of mechatronics

Words and Expressions

mechatronics	[ˌmekə'trɒnɪks]	*n.* 机电一体化 , 机械电子学
mechanism	['mekənɪzəm]	*n.* [C] 机械装置；机械作用；机构；结构
mechanics	[mi'kænɪks]	*n.* 力学，机械学
electronics	[ɪ'lek'trɒnɪks]	*n.* 电子学
interdisciplinary	[ˌɪntə'dɪsəplɪnərɪ]	*adj.* 跨学科的 , 多领域的，各学科间的
modeling	['mɒdlɪŋ]	*n.* 建模，造型，模特儿职业
		adj. 制造模型的，模特儿的
sensor	[' sensə]	*n.* 传感器，灵敏元件
automotive	[ˌɔ:tə'məʊtɪv]	*adj.* 自动的，汽车的
automotive vehicles		机动车辆
optoelectronic	[ˌɒptəʊɪlek'trɒnɪk]	*adj.* 光电（子）的，光电子学的，光电耦合的
opto-coupler		光电耦合
intelligent	[ɪn'telɪdʒənt]	*adj.* 聪明的，有智力的，智能的；了解的，熟悉的 [(+*of/about*)]
forward-thinking		高瞻远瞩的，有远见的
actuator	['æktʃʊeɪtə]	*n.* 调节器，激励器，执行元件
robotics	[rəʊ'bɒtɪks]	*n.* （用作单数）机器人学，机器人技术
robot	['rəubɒt]	*n.* [C] 机器人；自动控制装置；遥控装置
encompass	[ɪn'kʌmpəs]	*n.* 围绕，包含 过去式：encompassed 过去分词：encompassed 现在分词：encompassing
transaction	[træn'zækʃn]	*n.* 事项，记录，处理，交易 [C]；（pl）学报，会刊
synergistic	['sɪnədʒɪstɪk]	*adj.* 协作的，合作的，互相促进的
integration	[ˌɪntɪ'greɪʃn]	*n.* 整体化，集成，综合 , 混合，融合
category	['kætəgərɪ]	*n.* 种类，目录，部门
fusion	['fju:ʒn]	*n.* 融合，联合，聚变
fuse	[fju:z]	*n.* 保险丝
mechanical	[mə'kænɪkl]	*adj.* 机械的，力学的，机械般的
discipline	['dɪsəplɪn]	*n.* 纪律，训练 *vt.* 训练，处罚，自我控制 过去式：disciplined 过去分词：disciplined 现在分词：disciplining

| legitimacy | [lɪ'dʒɪtɪməsɪ] | *n.* 合法（性）；正统（性）；合理 |
| premier | ['premɪə(r)] | *n.* 首相，总理　*adj.* 首位的；首要的；最早的，最先的 |

IEEE *abbr.* Institute of Electrical and Electronic Engineers
　　　　电气和电子工程师协会

ASME *abbr.* American Society of Mechanical Engineers
　　　　美国机械工程师学会

latest	['leɪtɪst]	*adj.* 最新的，最近的；最迟的
newly	['nju:lɪ]	*adv.* 新近，最近；重新，再次
adaptable	[ə'dæptəbl]	*adj.* 能适应新环境的；适应性强的；适合的
journal	['dʒɜ:nəl]	*n.* [C] 日报；杂志；期刊；日记；日志
micro	['maɪkrəu]	*abbr.* microcomputer/microprocessor 前缀 *pref.* ①表示"小""微" ②表示"缩微的" ③表示"扩大的"
macro	['mækrəu]	*adj.* 巨大的，大量的；宏观的　*n.* 巨（计算机术语）；宏指令 前缀 *pref.* ①表示"大的""长的" ②表示"宏观的"
at least		至少，最低限度
in part		在某种程度上；部分地
coin	[kɔɪn]	*n.* 硬币，钱币　*vt.* 铸造（货币）；创造，杜撰（新词等）
array	[ə'reɪ]	*n.* （军队等的）列阵；（排列整齐的）一批；一系列；大量 [(+*of*)]

Exercise 1: Special terms

1. mechatronics　　　　＿＿＿＿＿＿＿＿＿＿＿

2. mechanics　　　　　＿＿＿＿＿＿＿＿＿＿＿

3. electronics　　　　　＿＿＿＿·＿＿＿＿＿＿

4. 跨学科的，多领域的，各学科间的　＿＿＿＿＿＿＿＿＿＿＿

5. 光电（子）的，光电子学的，光电耦合的　＿＿＿＿＿＿＿＿＿＿＿

6. 调节器，激励器，执行元件　＿＿＿＿＿＿＿＿＿＿＿

Exercise 2: Answer the following questions.

1. Is the mechatronics something new?

2. Who invented the term "mechatronics"?

3. When did mechatronics gain legitimacy in academic circles?

4. There are several topics that can be included in the category of mechatronics, what are they?

Exercise 3: Define the following terms with information from Text A.

1. mechatronics

2. mechatronics (until the early 1980s)

3. mechatronics (in the mid-1980s)

Part Two: Situational Conversation

Topic introduction: Tom and Mary have now just finished a class given by Prof. Smith on the work of the machine tools. Yet, they still have some questions, and so, they walked over to Prof. Smith.

Tom: (to Prof. Smith) Excuse me, Prof. Smith, I'm Tom, one of your students in this class. Could I ask you some questions?

Prof. Smith: (to Tom) Why not? Come on, then.

Tom: Why do we name these machines numerical control machines?

Tom: A good question, boy. You know, uh-hm…, NC, or numerical control, actually refers to the control of a machine tool or any other processing machines by using a series of mathematical information, or numerical data. It means the work of machines controlled by a numerical control program.

Mary: Oh, Prof. Smith, the idea is nice enough. But what is the advantage of numerical control machining over the hand control? Isn't our hand more sensible or reliable than the programmed data?

Prof. Smith: In some ways, yes. A good example is some fine art works, like a jade box, but the qualities of handmade products may not be consistent or stable. What is more important, NC has proved to be much more advantageous in overall operation.

Tom: Thank you, Prof. Smith, but I have another question. How can we use NC to get more satisfactory results in the real production?

Prof. Smith: It's better to use NC tools together with other technical advances, such as programmed optimization of cutting speed and feeds, work position, tool selection, and chip disposal.

Tom & Mary: Thank you Prof. Smith, NC sounds really promising from what you say.

Notes:

1. optimization 优化
2. cutting speed 切削速度
3. cutting feeds 切削进给
4. work position 工件定位
5. chip disposal 切屑处理

新编机电英语

Exercise 1: Read the following dialogues and fill the blanks with the suggested expressions.

Suggested expressions:

1. you've listened above
2. mechatronically
3. in many ways

Tom: Mary, I just wonder what's the difference between mechanical products and mechatronical products.

Mary: Uh, let me think how to express myself clearly. Actually, the difference can be made clear_____. Let's take washing machines, electronic cooking appliances, or, office equipment-fax, plain paper copiers, for example. How can you make such things with purely mechanical method?

Tom: Yes, indeed. But things_____still belong to mechanical product, don't they?

Mary: Well, I think in some ways they are. But remember all of them are made _____, rather than mechanically.

Exercise 2: You're required to compare and find some differences between the products made mechanically and mechatronically.

1. Products to be compared: A TV receiver, a modern printer and an old fashioned typewriter.

2. Aspects to be compared: flexibility; precision; outlook; material used; ways of production.

⚑ Part Three: Reading Comprehension

Despite the brouhaha(骚动)over stolen e-mails from the University of East Anglia, the science of climate change is so well established enough by now that we can move on the essential question: what's the damage going to be?

The total bill, if emissions are left unchecked, could reach 20 percent of annual per capita income, says Nicholas Stern, the British economist who led an influential Whitehall-sponsored study. William Nordhaus, a Yale economist, puts his "best guess" at 2. 5 percent of global yearly GDP. And according to Dutch economist Richard Tol, the economic impact of a century's worth of climate change is "relatively small" and "comparable to the impact of one or two years of economic growth".

These estimates aren't just different—they're different by an order of magnitude. And while some might dismiss the cost estimates as mere intellectual exercises, they're intellectual exercises with real impact. The Copenhagen Meeting may be a bust, but countries from the United States to China are individually considering cap-and-trade schemes, carbon taxes, and other policies aiming at curtailing greenhouse gases. To be effective, a tax or cap—and trade charge would have to force today's emitters to pay the true "social cost of carbon"—in other words, the amount of damage a ton of carbon will cause in the coming centuries.

Figuring out what that cost is, however, is not a simple task. That's largely because most of the bill won't come due for many decades. A ton of carbon dioxide emitted today will linger in the air for anywhere from one to five centuries. Virtually every cost study shows that, even if economic growth continues apace（快速地）and there's no effort to slash emissions, the damage from climate change will be negligible until at least 2075. It could take 100 years before we see noticeably negative effects, and even more before we need to launch massive construction projects to mitigate（减轻）the damage.

1. What can we learn from the first paragraph?

    A) Those stolen e-mails should not be made fuss of.

    B) It is time to talk about damage caused by climate change.

    C) The science of climate should have been established.

    D) Climate change is essential to human beings.

2. What's the damage brought by climate change according to the author?

    A) It account for 20% of annual per capita income.

    B) It is about 2.5% of the yearly global GDP.

    C) Its damage is overestimated by many economists.

    D) Economists haven't reached consensus yet.

3. What should be done to reduce carbon emissions?

    A) Gas emitters should pay for the damage.

    B) Policies should aim at reducing carbon emission.

C) Social cost of carbon should be shown to the public.

D) The damage should not be neglected.

4. Why is it hard to figure out the social cost of carbon?

A) Too many factors should be taken into account.

B) There is no effort aiming at carbon emissions reduction.

C) The damage cannot be seen until years later.

D)The damage will last for years before eliminated.

5. What can we learn about the present climate change?

A) The damage is somewhat exaggerated.

B) Action will be taken the moment people realize it.

C) Measures should be taken immediately to tackle it.

D) The negative effect will not be significant for a century.

Part Four: Translation Skills

拆句法和合并法

拆句法和合并法是两种相对应的翻译方法。拆句法是把一个长而复杂的句子拆译成若干个较短、较简单的句子，通常用于英译汉；合并法是把若干个短句合并成一个长句，一般用于汉译英。汉语强调意合，结构较松散，因此简单句较多；英语强调形合，结构较严密，因此长句较多。所以汉译英时要根据需要注意利用连词、分词、介词、不定式、定语从句、独立结构等把汉语短句连成长句；而英译汉时又常常要在原句的关系代词、关系副词、主谓连接处、并列或转折连接处、后续成分与主体的连接处以及意群结束处将长句切断，译成汉语分句。这样就可以基本保留英语语序，顺译全句，顺应现代汉语长短句交替、单复句相间的句法修辞原则。

例1： Increased cooperation with China is in the interests of the United States.

翻译：同中国加强合作，符合美国的利益。（在主谓连接处拆译）

详解：根据意义上的需要，一个复杂的句子可以拆分为两个小句。increased 的词根（increase）本就有动词词性的意思，所以把 increased cooperation with China，翻译成与中国加强合作合乎情理，句子也短而精练，be in the interests of 表示与"与……利益相符"，把 be 翻译成"符合"，而非"是"，符合汉语的词组习惯。

例2： I wish to thank you for the incomparable hospitality for which the Chinese people are justly famous throughout the world.

翻译：我要感谢你们无与伦比的盛情款待。中国人民正是因为这种热情好客而闻名世界的。（在定语从句前拆译）

详解：从 for which 这个结构可以看出后半句 the Chinese people are justly famous throughout the world 是一个定语从句。定语从句是一个复杂句，此时就可以用拆句法，把定语从句翻译成一个小句，其主句也可以翻译成一个小句。通过"for which"可以判断两个小句的关系是表原因，所以前一句翻译成"我要感谢你们无与伦比的盛情款待"，后一小句翻译成"中国人民正是因为这种热情好客而闻名于世的"，符合汉语语言规范。

例 3：中国是一个发展中的沿海大国。中国高度重视海洋的开发和保护，把发展海洋事业作为国家发展战略。

翻 译：As a major developing country with a long coastline, China attaches great importance to marine development and protection, and takes it as the state's development strategy.（合并法）

详解：此处原句被翻译成一个长长的英语单句。原句的第一句被译成一个由 as 引导的介词短语，表示作为一个发展中的沿海大国，主语仍然是中国，主管两个并列动词："高度重视"和"把……作为"，其对应的英文词组就是 attach importance to 和 take or regard...as，符合英语的惯用语，也表达了汉语的基本意思。

Translation Exercise 1

机电一体化技术并不是什么新事物，它是指把精密机械工程、控制理论、计算机技术和电子技术等领域的最新技术结合起来应用于设计过程中，以创造出功能更强、适应性更好的产品。当然，这也是很多具有超前思想的设计师和工程师多年来一直致力于的事情。

Translation Exercise 2

常见故障及排除方法：如果故障是按动开关却没有任何声响，请检查是否：①已接电源；②选择了正确的输入；③音量开到最小；④音箱已正确接驳；⑤主机处于静音状态。

Unit 6 Electronic Communication

Learning Objectives

After completing this unit, you will be able to do the following:

1. Grasp the main idea and the structure of the text;
2. Master the key language points and grammatical structure in the text;
3. Understand the basic materials used to make semi-conductor;
4. Conduct a series of reading, listening, speaking and writing activities related to the theme of the unit.

Outline

The followings are the main sections of this unit.

1. Warm-up Activity;
2. Text A;
3. Situational Conversation;
4. Reading Comprehension;
5. Translation Skills;
6. Exercises.

Electronic Communication Terms

In this unit, you will learn some technical terms in electronic communication listed below.

• silicon;
• transistor;
• integrated circuit;
• electric current.

Vocabulary

The listed below are some words appearing in this unit that you should make part of your vocabulary.

• oxygen;
• atom;
• refine;
• electron;
• vacuum tube.

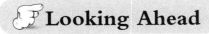

Looking Ahead

Semi-conductor

As one of the most important parts in the area of electronic communication, semi-conductors are widely used in various electronic communication devices. A semi-conductor is a material with conducting properties between those of a good insulator like glass and a good conductor like copper. The most commonly used semi-conductor is silicon. Each silicon atom has an outer shell with four valence electrons and four vacancies. In pure silicon, atoms join together by forming covalent bonds. Each atom share its valence electrons with each of four adjacent neighbours effectively filling its outer shell. The structure of pure silicon has zero overall charge. The complete nature of the structure means that at absolute zero temperature none of the electrons are available for conduction, thus the material is an insulator.

Core Contents

Pure silicon is an insulator with great resistance. This is because the silicon has four electrons, and each silicon atom is connected to four other silicon atoms, which is like carbon atoms in diamond, in which the electrons mostly have got into pairs, through which the current does not easily flow. After adding a further element, such as phosphorus which has five electrons outside. So, the phosphorus atom combined with four silicon atoms, leaving one more electron, which can make movement in the silicon crystal and generate a current. Now the resistance value of pure silicon is decreased between the conductor and insulator, so it is called semi-conductor.

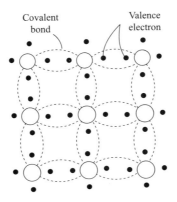

Covalent bond   Valence electron

Warm-up Activity

What are widely used in making electronic communication devices?

condenser   chip

Part One: Text A What Is Semi-conductor?

Silicon, taking up about 28% of the elements on the earth, is one of the largest amount of elements besides oxygen on the surface of the Earth. There are no pure silicon in nature, and it usually exists as compounds, such as silicon dioxide in sand and silicate in glass, but they are all not pure. Silicon was used to be thought useless, but later, it was found of the semi-conductor electrical characteristics, which made it into a very important element. It can be made into transistors, with the feature of amplifying the electronic signals, which is the working principle of radio. Vacuums were used in electrical signal amplification at beginning. A vacuum tube is about 2 cm in diameter and 5-6 cm high. Because a vacuum tube is very large, the size of a radio before was as large as a modern TV set. The computer was first invented with vacuums dealing with electric signals. It took about one room to put it in and needed a lot of electricity to operate it. After the invention of transistor, it replaced the vacuum tube's work to amplify the electronic signals. In the size of only about a watermelon seed, many appliances using vacuums before have been transformed into using of transistors. Therefore, the radio now is only as big as a pack of cigarettes. Now a lot more transistors are put together with all of its production lines becoming integrated circuits. A fingernail-sized piece of an integrated circuit is equivalent to tens of thousands or even more transistors.

The common silicon, namely silicon dioxide, is dark brown, as there are a lot of impurities in it. Relatively pure sand is white, and pure silicon is available in this white sand with carbon reduction. Pure silicon is dark brown with a metallic sheen. The purity of the silicon produced through this chemical method can only reach about 98%, which is still not pure enough for electronic industry. Then using the conventional distillation process in industry is a better way to get more pure silicon. At last, making use of the regional refining method, silicon can be purified to the extent of about only one impurity atom in every a hundred million silicon atoms and even purer, meanwhile, crystalline silicon, which is also called silicon rod, is also made in this process. After slicing, it becomes what is commonly known as a silicon chip. There was no practical value of the initially got silicon rod, as its diameter was too small. The silicon rods didn't have any practical value until they could get to three inches or four. At present, twelve inches or larger silicon rods can be made.

Pure silicon is an insulator with great resistance. This is because the silicon has four electrons, and each silicon atom is connected to four other silicon atoms, which is like carbon atoms in diamond, in which the electrons mostly have got into pairs, through which the current does not easily flow. After adding a further element, such as phosphorus which has five electrons outside. So, the phosphorus atom combined with four silicon atoms, leaving one more electron, which can make movement in the silicon crystal and generate a current. Now the resistance value of pure silicon decreases between the conductor and insulator, so it is called semi-conductors, also known as N-type semi-conductor. Similarly, there are other elements such as boron, with only three electrons outside its atom. When a boron atom combines with four silicon atoms, an electron lacks, so, an electrical hole is created, because an electron can be filled in it. This hole is equivalent to a positive charge, which can also make a movement in the silicon crystal and generate a current. This is known as P-type semi-conductor. Adding this kind of impurity is called doping. The silicon of very high purity is needed as the amount of these impurities have to be controlled in a very low level. When a P-type semi-conductor is linked to an N-type semi-conductor, it is called PN junction. A PN junction is characterized by rectification, in which the current flows in only one direction. If other electrodes are added, it can handle electrical signals, which is a transistor. The silicon after reduction can be made into solar batteries, but the current efficiency is not very good. Silicon can also be made into a lot of very useful compounds, such as silicon carbide, which is a very hard material, or glass, which is a very commonly used material.

Words and Expressions

silicon	['sɪlɪkən]	n. [化学] 硅；硅元素
oxygen	['ɒksɪdʒən]	n. [化学] 氧气，[化学] 氧
compound	['kɒmpaʊnd]	n. [化学] 化合物
dioxide	[daɪ'ɒksaɪd]	n. 二氧化物
silicate	['sɪlɪket]	n. [矿物] 硅酸盐
transistor	[træn'zɪstɚ]	n. 晶体管

carbon	['kɑrbən]	*n.* 碳
deacidize	[diːˈæsɪdaɪz]	*vi.* 脱氧；还原
purity	['pjʊərɪtɪ]	*n.* ［化学］纯度
distill	[dɪsˈtɪl]	*vt.* 提取；蒸馏；使滴下　*vi.* 蒸馏；滴下；作为精华产生（等于 distil）
atom	['ætəm]	*n.* 原子
refine	[rɪˈfaɪn]	*vt.* 精炼，提纯
crystal	['krɪst(ə)l]	*n.* 结晶，晶体
silicon rod		硅棒
insulator	['ɪnsjʊleɪtə]	*n.* ［物］绝缘体
electron	[ɪˈlektrəʊn]	*n.* 电子
phosphorus	['fɒsf(ə)rəs]	*n.* 磷
boron	['bɔːrɒn]	*n.* ［化学］硼
impurity	[ɪmˈpjʊərɪtɪ]	*n.* 杂质
mix	[mɪks]	*vt.* 配制；混淆；使混和　*vi.* 参与；相混合
rectification	[ˌrektɪfɪˈkeɪʃən]	*n.* ［电］整流
electrode	[ɪˈlektrəʊd]	*n.* ［电］电极
silicon dioxide		二氧化硅
vacuum tube		真空管
electricity signal		电讯号
integrated circuit		集成电路
positive charge		正电荷
electric current		电流

Exercise 1: Special terms.

1. junction　　　　　　　_____
2. generate a current　　_____
3. electronic signal　　　_____
4. 接线盒　　　　　　　_____
5. 风力发电　　　　　　_____
6. 数字信号　　　　　　_____

Exercise 2: Answer the following questions.

1. Which element constitutes a semi-conductor?

2. What's the difference in colour between common silicon and pure silicon?

3. How to purify silicon in order to achieve the standard of electronic industry?

4. What are the other aspects that silicon can be applied to besides making semi-conductors?

Exercise 3: Define the following terms with information from Text A.

1. semi-conductor

2. silicon chip

3. transistor

Part Two : Situational Conversation

(There are two colleagues in a tech company)

A: How do you plan to distribute the information about last week's meeting?

B: I am planning on sending a bulk email to all the users on our company's server.

A: That's a time-saving way to get the word out, but I don't think it will be that effective. Most of us just delete the bulk emails without even reading them. Even if you put a really catchy（容易记住的）subject line in there, I don't think anyone will get the information.

B: Well, it would be too much time-consuming to send the emails one by one to our entire staff… What do you suggest?

A: You could send the meeting brief to the managers by email, and ask them to forward it to their subordinates. Most people will read an email if it is sent by their supervisor.

B: That's a good idea. I can just put the meeting minutes on the email as an attachment, then forward it along to the managers. Can you show me how to make an attachment with our email program?

A: I'm sorry, I know next to nothing about the new email program. It's supposed to be more user-friendly than our last program, but I still haven't figured it out.

B: I'll ask our tech support for some help. It shouldn't be too complicated, I'm sure.

Notes:

1. "I am planning on sending a bulk email to all the users on our company's server." 我计划向我们公司服务器上的所有用户发群邮件。 "plan on doing sth." 计划做什么事情； "bulk" 大量，这里 "bulk email" 指群邮件、垃圾邮件； "server" 指服务器。

2. get the word out：广而告之，宣传。

3. subject line：主题大纲。

4. "You could send the meeting brief to the managers by email, and ask them to

forward it to the people underneath them." 你可以先把会议提要通过邮件发给各部门经理，再让经理转发给他们的下属。"forward" 在这里做及物动词，表示"转寄、转送、转发"。

5 "I can just put the meeting minutes on the email as an attachment." 我可以把会议纪要上传至附件中发送。"meeting minutes" 指会议记录，会议纪要；"attachment" 指附件。

6. "Can you show me how to make an attachment with our email program?" 你知道用我们新的邮件程序如何添加附件吗？"program" 指程序。

7. "I know next to nothing about the new email program." "know next to nothing about" 指几乎一无所知。

8. user-friendly：容易使用的。

9. figured it out：搞清楚，弄明白。

Exercise 1: Sentence patterns.

1. I am planning on sending a bulk email to all the users on our company's server.

2. That's a time-saving way to get the word out.

3. Most of us just delete the bulk emails without even reading them.

4. You could send the meeting brief to the managers by email, and ask them to forward it to their subordinates.

5. I can just put the meeting minutes on the email as an attachment, then forward it along to the managers.

6. I know next to nothing about the new email program.

7. …but I still haven't figured it out…

Exercise 2: Complete the following dialogue in English.

(There are two company employees.)

A: Let's correspond by email. _____.
（我认为我们应该保持联系）There are a lot of future opportunities to work together.

B: I agree. _____. （我对你所说的 Aluminum 项目尤其感兴趣）_____? （你能把细节发送给我吗？）

A: Sure!_____. （我一回到办公室就一并发给您）
I did get your card, didn't I?

B: Oh, I almost forget! Here it is.

A: Thank you. Is all the information on here current?

B: Let's see… Yes, but there is only my work email address. I'll give you my personal address too. _____. （有时候如果附件过大，我的工作邮箱

就会拒收）_____.（如果你的附件超过 100KB，请发至我的个人邮箱）

A: It'll be alright. _____.（我可以发一个压缩文件）Does your computer have the software to unzip files?

B: I do, but unzipping files doesn't work out so well. _____.（上一次我试着解压文件，整个电脑系统都崩溃了）If it is a large file, it would probably work better to send it to my personal email. It's better to be safe than sorry.

A: No problem, I will make sure to email the information to your personal email address first thing.

Part Three: Reading Comprehension

British psychologists have found evidence of a link between excessive Internet use and depression, a research has shown.

Leeds University researchers, writing in the Psychopathology journal, said a small proportion of Internet users were classed as Internet addicts and that people in this group were more likely to be depressed than non-addicted users.

The article on the relationship between excessive Internet use and depression, a questionnaire-based study of 1,319 young people and adults, used date gathered from respondents to links placed on UK-based social networking sites.

The respondents answered questions about how much time they spent on the Internet and what they used it for; they also completed the Beck Depression Inventory—a series of questions designed to measure the severity of depression.

The six-page report, by the university's Institute of Psychological Sciences, said 18 of the subjects who completed the questionnaire were Internet addicts.

"Our research indicates that excessive Internet use is associated with depression, but what we don't know is which comes first do depressed people drawn to the Internet cause depression?"

The article's lead author, Dr. Catriona Morrison, said. "What is clear is that, for a small part of people, excessive use of the internet could be a warning signal for depressive tendencies."

The age range of all respondents was between 16 and 51 years old, with a mean age of 21.24. The mean age of the 18 internet addicts, 13 of whom were male and five female, was 18.3 years. By comparing the scale of depression within this group to that within a group of 18 non-addicted internet users, researchers found the internet addicts had a higher incidence of moderate to severe depression than non-addicts. They also discovered that addicts spent proportionately more time browsing sexually pleasing websites, online gaming sites and online communities.

"This study reinforces the public speculation（推测）that over-engaging in websites that serve to replace normal social function might be linked to psychological disorders like depression and addiction," Morrison said. "We now need to consider the wider societal implications of this relationship and establish clearly the effects of excessive internet use on mental health."

1. internet addicts are people who _____.

    A) use the internet more than enough

    B) feel depressed when using the internet

    C) seldom connect to the internet

    D) feel depressed without the internet

2. What was collected as data by the researchers?

    A) The number of users of UK-based social networking sites.

    B) Respondents' visits to UK-based social networking sites.

    C) Links on UK-based social networking sites.

    D) Respondents' answers to a questionnaire.

3. What is confirmed by the study?

    A) Depression leads to excessive use of internet.

    B) Depression results from excessive use of internet.

    C) Excessive use of internet usually accompanies depression.

    D) Excessive use of internet is usually prior to depression.

4. It is speculated by the public that online communities _____.

    A) can never replace normal social function

    B) are intended to replace normal social function

    C) are associated with psychological disorders

    D) shouldn't take the blame for psychological disorders

5. According to Dr. Catriona Morrison, the public speculation _____.

    A) lacks scientific evidence   B) helps clarify their study

    C) turns out to be correct   D) is worth further study

Part Four: Translation Skills

正译法和反译法

正译法和反译法通常用于汉译英，偶尔也用于英译汉。所谓正译，是指把句子按照与原语相同的语序或表达方式译入目的语。所谓反译则是指把句子按照与原语相反的语序或表达方式译入目的语。正译和反译常常具有同义的效果，但反译往往符合英语的思维方式和表达习惯，因此比较地道。

例 1： 在美国，人人都能买到枪。

正译： In the United States, everyone can buy a gun.

反译： In the United States, guns are available to everyone.

详解： 反译体现出英语更倾向于用被动句表示句子。人买枪，符合汉语习惯，但枪能被大多数人所获得，则符合英语逻辑。available 就表示可获得的意思。

例 2： 他突然想到了一个新主意。

正译： Suddenly he had a new idea.

He suddenly thought out a new idea.

反译： A new idea suddenly occurred to him.

详解： 这里体现了英语的惯用语序。他想到一个主意，符合汉语习惯，但一个主意突然迸发到人的脑海中，符合英语逻辑。A new idea suddenly occurred to him. 这个翻译更地道。sth. occurs to sb. 是英文的地道用法。

例 3： 你可以从因特网上获得这一信息。

正译： You can obtain this information on the Internet.

反译： This information is accessible/available on the Internet.

详解： 这里体现了英语更倾向于用被动句表示句子的含义。你从因特网获得信息，符合汉语习惯，但信息在因特网上可获取或获得，则符合英语逻辑。available 或 accessible 表示可获得或是通过某种途径可以接触到的意思，词和中文一一对应，且符合英语说法，更显地道。

Translation Exercise 1

感谢您使用本公司出品的数码产品，为了让您轻松体验产品，我们随机配备了内容详尽的使用说明，您从中可以获取有关产品的介绍及使用方法等方面的知识。在您开始使用本机之前请先仔细阅读说明书，以便您能正确地使用本机，如有任何印刷错误或翻译失误望广大用户谅解，当涉及内容有所更改时，恕不另行通知。

本机是一款外观小巧、设计精美、携带方便的多媒体小音响，适用于家居、户外、办公室等场所，可以随时随地享受音乐带来的轻松，为您的电脑、数码音乐播放器、手机等视听产品提供超值完美的音质。

Translation Exercise 2

本机的微电脑系统自动检测识别外接设备，开机后进入待机模式，插入 U 盘或 TF 储存卡自动识别播放，插入音频信号线自动识别播放，后者优先原则，详细功能操作请阅读第四项"产品的按键、插孔功能定义"。

将电源线一端插入本机的 DC 5 V 插孔，另一端 USB 插头插入 PC 的 USB 接口，或连接标准 5V 500mA 的充电器接口，充电中"紫色灯"闪动，充满电量后"紫色灯"停止闪动。

Unit 7 Information Technology

Learning Objectives

After completing this unit, you will be able to do the following:

1. Grasp the main idea and the structure of the text;
2. Master the key language points and grammatical structure in the text;
3. Understand different aspects in which the 3G is applied in our life;
4. Conduct a series of reading, listening, speaking and writing activities related to the theme of the unit.

Aviation Terms

In this unit, you will learn some technical terms in information technology listed below:

- wideband;
- capacity;
- access speeds.

Outline

The followings are the main sections of this unit:

1. Warm-up Activitiy
2. Text A;
3. Situational Conversation;
4. Reading Comprehension;
5. Translation Skills;
6. Exercises.

Vocabulary

The listed below are some words appearing in this unit that you should make part of your vocabulary:

- stereo;
- intensify;
- boost;
- log on;
- vacancies;
- wireless network;
- maintenance engineer.

Looking Ahead

3G Technology

Mobile communications in the 21st century will enter an era of mobile multimedia service and universal mobility. Various technological developments are being carried forward towards this goal, including the development of an advanced intelligent network that will integrate different communications systems to establish a sophisticated mobile communications network which will realize these services. The global standard for third-generation wireless communications has been determined by the International Telecommunication Union (ITU) under the integrated name of IMT-2000. IMT-2000 realizes mobile communications systems that offer high quality equivalent to that of fixed networks under global standard radio interfaces and can provide a wide range of services. In addition to making it possible to easily communicate with anybody from anywhere at anytime on a global scale, it also permits high-speed, large-volume data communications and image transmissions.

Core Contents

1. The definition of 3G technology.
2. What are the effects about 3G technology?

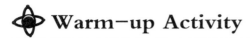 **Warm-up Activity**

What are the important factors in developing an information society?

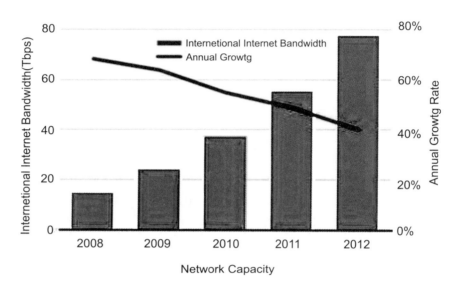

Internetional Internet Bandwidth Gowth, 2008-2012

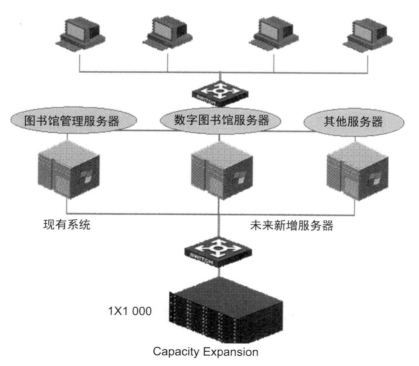

Part One: Text A The Application of 3G in Our Life

Although the technology behind 3G may seem complicated, the ways in which 3G will affect all of our lives are easy to imagine. Just imagine having a combined camera, video camera, computer, stereo, and radio included in your mobile phone. Rich-media information and entertainment will be at your fingertips whenever you want anywhere there is a wireless network.

Mobile communication is moving from simple voice to rich media, where we use more of our senses to intensify our experiences.

But not all of this will happen at once. 3G is an evolution to a communications ideal that no one completely understands yet. What we do know is that mobile multimedia will hit the Japanese markets in 2001, and Europe and North America will follow soon after.

3G brings together high-speed radio access and IP-based services into one powerful environment. The step towards IP is vital. IP is packet-based, which in simple terms, means users can be "on-line" all the time, but without having to pay until we actually send or receive data. The connectionless nature of IP also makes access a lot faster: file downloads can take a few seconds and we can be connected to our corporate network with a single click.

3G introduces wideband radio communications, with access speeds of up to 2Mbit/s. Compared with today's mobile networks, 3G will significantly boost network capacity—so operators will be able to support more users, as well as offer more sophisticated services.

3G at home

3G is going to affect our home and social lives in many ways. The services that 3G enables will help us to manage our personal information, simplify tasks such as grocery shopping, make better use of our time and offer services that are just fun to use. Operators will be able to develop myriad new service opportunities to attract and retain new customers. Here are some examples:

You're sitting on a train and use this "dead" time to log on to your bank account, check your balance and pay a few bills—all through your 3G device. You save time and can be smarter about managing your finances.

On a touring vacation, you arrive in a new city. You haven't made any reservations in advance, because you can do this when you get there, by using your 3G handset to obtain up-to-date information, including hotel vacancies. Having booked a room, you can use your mobile to view video clips of local tourist attractions and talk to someone from the local tourist information bureau at the same time.

3G at work

3G will not only support the needs of business people who travel a lot, but will also help new, flexible working practices, such as home-working and remote access to corporate networks outside traditional working hours. Business people are often high-volume airtime users, so they represent a big opportunity for mobile operators. Here are some examples:

At work you receive a message from your "smart" refrigerator at home. The message tells you that certain items need restocking and an order has already been prepared for the local grocery store, which you can approve, so that your groceries are ready to collect on the way home.

You are on the road, and urgently need to discuss a draft presentation with a number of colleagues back in the office. Pulling into a service station, you use your 3G device to hold a telemeeting with your colleagues and, at the same time, you can all view the draft presentation and make changes on line.

A maintenance engineer is repairing some equipment on a client's premises and hits a problem. Using his 3G device, he contacts his department and downloads a demonstration video that guides him through the repair process.

Words and Expressions

stereo	['sterɪəʊ]	*n.* 立体声；立体声系统
intensify	[ɪn'tensɪfaɪ]	*vt.* 使加强，使强化
packet		*n.* 数据包，信息包
connectionless		*adj.* 无链接的
corporate	['kɔ:p(ə)rət]	*adj.* 共同的，全体的；社团的
click	[klɪk]	*n.* 单击；滴答声
wideband	['waɪd'bænd]	*adj.* 可用多种频率的，宽频带的
boost	[bu:st]	*vt.* 促进；增加
capacity	[kə'pæsɪtɪ]	*n.* 能力；容量
grocery	['grəʊs(ə)rɪ]	*n.* 食品杂货店
myriad	['mɪrɪəd]	*adj.* 无数的；种种的
log on		登录；注册
handset		*n.* 手机

vacancies	['veɪk(ə)nsɪ]	*n.* 空缺；空位；空白；空虚
high-volume		*adj.* 大容量
smart	[smɑ:t]	*adj.* 智能的
restock	[ˌrɪ'stɑk]	*vt.* 重新进货；再储存　*vi.* 补充货源；补足
premise	['premɪs]	*n.* 前提；上述各项；房屋连地基
at one's fingertips		触手可及
wireless network		无线网络
radio access		无线接入
on line		在线
hit a problem		点击一个问题
maintenance engineer		维修工程师
access speeds		［计］存取速度；选取速度；接取速度

Exercise 1: Special terms.

1. wireless network　　＿＿＿＿＿＿＿＿＿＿＿＿
2. radio access　　＿＿＿＿＿＿＿＿＿＿＿＿
3. access speed　　＿＿＿＿＿＿＿＿＿＿＿＿
4. 无线通信　　＿＿＿＿＿＿＿＿＿＿＿＿
5. 无线宽带接入　　＿＿＿＿＿＿＿＿＿＿＿＿
6. 智能冰箱　　＿＿＿＿＿＿＿＿＿＿＿＿

Exercise 2: Answer the following questions.

1. What diretion is mobile communication moving to?
2. What are the advantages of 3G compared with 2G?
3. What is 3G applied for at home?
4. What is 3G applied for at work?

Exercise 3: Define the following terms with information from Text A.

1. IP
2. 3G

Part Two : Situational Conversation

(There are two colleagues discussing about researching and developing a new product.)

A: Hi, good afternoon. How are you?

B: I'm fine, recently, what are you doing?

A: Our company is researching and developing a news product!

B: Wow! It's cool. How do you develop a new product?

A: Well, first of all, of course, we need an idea.

B: Where are they from?

A: Well, from many sources.

B: For example?

A: Obviously, many ideas come from our own research and development department.

B: It's a good idea. They fought in R&D in the first line. So it's very important. What else?

A: We also get ideas from other employees in the company from our customers, from focus groups, from visiting trade shows.

B: The competitors knew you can win.

A: We also use more formal idea generating techniques.

B: So you've got an idea. How do you test techniques?

A: First, we ask ourselves key questions about the ideas. Next, make product test again and again until the product success rate reaches 100%.

B: That sounds great. I believe you will succeed.

A: Thank you.

B: See you.

Notes:

1. research and development department: 研发部门。

2. "They fought in R&D in the first line."

他们战斗在第一线（in the first line：第一线）。

3. focus group: 为测试民意而选定的典型群众。

4. visiting trade shows: 参观展会。

5. "idea generating techniques"：激发想法技术（generate：生成，产生）。

Exercise 1: Sentence patterns.

1. Our company is researching and developing a new product!

2. They fought in R&D in the first line.

3. The competitors knew you can win.

4. Next, make product test again and again until the product success rate reaches 100%.

Exercise 2: Complete the following dialogue in English.

(There are two colleagues discussing about a project leader.)

A: Hi, good afternoon. How are you?

B: It's OK. _____.(我听说你最近要做一个项目)

A: Yes. But we haven't a project leader. So, we need to choose one.

B: _____?（你需要项目领导者具备哪些良好的素质）

A: Basically, _____. （我们需要一个人能够与团队的其他成员协同一致）

B: It's very nice. In my opinion, a leader can bring us a hamonious environment. What else?

A: Yes, _____.（但是这个人也应该能够激励和领导大家）

B: Yes. It is good for team, and it is very important to the progress. What's your opinion?

A: _____. （一个好的项目领导者必须能在更大的组织面前代表他们团队的利益）They should be a sort of champion for the project.

B: I agree it. That's great and then what?

A: At the same time. _____.（我们需要一个有责任感的人）

B: Yes. You're right. _____.（正是项目领导人把项目提上日程并且按照预算来做）

A: Thank you!

B: You're welcome. See you.

⚑ Part Three: Reading Comprehension

If you're like most people, you're way too smart for advertising. You skip right past newspaper ads, never click on ads online and leave the room during TV commercials.

That, at least, is what we tell ourselves. But what we tell ourselves is wrong. Advertising works, which is why, even in hard economic times, Madison Avenue is a $34 billion-a-year business. And if Martin Lindstrom—author of the best seller Biology and a marketing consultant for Fortune 500 companies, including PepsiCo and Disney—is correct, trying to tune this stuff out is about to get a whole lot harder.

Lindstrom is a practitioner of neuromarketing（神经营销学）research, in which consumers are exposed to ads while hooked up to machines that monitor brain activity, sweat responses and movement in facial muscles, all of which are markers of emotion. According to his studies, 83% of all forms of advertising principally engage only one of our senses: sight. Hearing, however, can be just as powerful, even though advertisers have taken only limited advantage of it. Historically, ads have relied on slogans to catch our ear, largely ignoring everyday sounds—a baby laughing and other noises our bodies can't help paying attention to. Weave this stuff into an ad campaign, and we may be powerless to resist it.

To figure out what most appeals to our ear, Lindstrom wired up his volunteers, then played them recordings of dozens of familiar sounds, from McDonald's wide-spread "I'm Lovin' It" slogan to cigarettes being lit. The sound that blew the doors off all the rest—both in terms of interest and positive feelings—was a baby giggling. The other high-ranking sounds were less original but still powerful. The sound of a vibrating cell phone was Lindstrom's second-place finisher. Others that followed were an ATM dispensing cash and a soda being popped and poured.

In all of these cases, it didn't take an advertiser to invent the sounds，combine them with meaning and then play them over and over until the subjects being part of them. Rather, the sounds already had meaning and thus triggered a series of reactions: hunger, thirst, happy anticipation.

1. As is mentioned in the first paragraph, most people believe that _____.

    A) Ads are a waste of time.   B) Ads are unavoidable in life.

    C) They are easily misled by ads.   D) They not influenced by ads.

2. What do we know about Madison Avenue in hard economic times?

    A) It becomes more thriving by advertising.

    B) It turns to advertising so as to survive

    C) It helps spread the influence of advertising

D) It keeps being prosperous thanks to advertising

3. What do we learn about PepsiCo and Disney form the passage?

A) Lindstrom was inspired by them to write a book.

B) They get marking advice from Lindstrom.

C) Lindstrom helps them to go through hard times.

D) They attribute their success to Lindstrom.

4. It is pointed out by Lindstrom that advertisers should _____.

A) rely more on everyday sounds

B) rely more on neuromarketing

C) rely less on slogans

D) rely less on sight effect

5. It is found by Lindstrom that a baby giggling is _____.

A) the most touching   B) the most familiar

C) the most impressive   D) the most distinctive

Part Four: Translation Skills

倒置法

  在汉语中，定语修饰语和状语修饰语往往位于被修饰语之前，而在英语中，许多修饰语则常常位于被修饰语之后，因此翻译时往往要把原文的语序颠倒过来。倒置法通常用于英译汉，即对英语长句按照汉语的习惯表达法进行前后调换，按意群或进行全部倒置，原则是使汉语译句符合现代汉语论理叙事的一般逻辑顺序。如：

例 1： At this moment, through the wonder of telecommunications, more people are seeing and hearing what we say than on any other occasions in the whole history of the world.

翻译： 此时此刻，通过现代通信手段的奇妙，能看到和听到我们讲话的人比史上这世界中任何其他场合的人都要多。（部分倒置）

详解： 根据意义上的需要，可以在主语和谓语之间进行倒装。如果把主语 more people 直接放在句首，例 1 中的 seeing and hearing what we say 的翻译似乎不够明确。通过主语和谓语的部分倒装，把谓语动词提到前面来，意思更为明确和清晰，读起来也较通顺自然，符合汉语习惯。

例 2： I believe strongly that it is in the interest of my countrymen that Britain should remain an active and energetic member of the European Community.

翻译： 我坚信，英国依然应该是欧共体中的一个积极的和充满活力的成员，这是符

合我国人民利益的。（部分倒置）

详解：根据意义上的需要，可以把两个从句进行倒装。如果把第一个从句 that it is in the interest of my countrymen 的翻译直接放在句首，则 that Britain should remain an active and energetic member of the European Community 的翻译似乎不够明确。通过把两个从句进行倒装，意思更为明确和清晰，读起来也较通顺自然。

例 3：Great changes have taken place in China since the introduction of the reform and opening policy.（全部倒置）

翻译：改革开放以来，中国发生了巨大的变化。

详解：根据意义上的需要，可以在主语和谓语之间进行倒装。如果把主语 Great changes 直接放在句首翻译，则 have taken place in China 的翻译似乎不够明确。通过主语和谓语的部分倒装，把谓语动词提到前面来，意思更为明确和清晰，读起来也较通顺自然，符合汉语习惯。

Translation Exercise 1

按键有几个简单按钮。"Power"键：使您可以轻松地开启或关掉手机。主屏幕按钮：可以让您随时退回到主屏幕界面。侧键：切换键，可以在菜单（返回）按钮和音量调节按钮之间切换。如果您当前没有使用手机，则可以锁定它以关闭显示屏，从而节省电池电量。如果已锁定手机，则触摸屏幕不起任何作用。手机仍可以接听电话，接收短信以及接收其他更新内容。您也可以听音乐并调整音量以及使用耳机上的中央按钮（或蓝牙耳机上的等效按钮）来播放或暂停播放歌曲，或者接听或结束通话。默认情况下，如果您不触摸屏幕的时间达到一分钟，则手机会锁定。

锁定需按下"Powerv 键按钮。

解锁需按下主屏幕按钮键或"Power"键点亮屏幕后拖移滑块实现解锁。

关机需按住"Power"键按钮几秒钟，直至出现手机选项——"关机"，选择"关机"，手机会关闭应用程序进入到关机状态。开机长按住"Power"键按钮几秒，直至出现开机标志。

Translation Exercise 2

当广播电台发射的电磁能量波扫过收音机或电视机接收器天线时，就会在天线里产生一股微弱的脉动电流，这种电流与其他能量波在同一天线中同时产生的感应电流发生电子脱离。这种选定电流（用于某一特定的收音波段或电视频道）被放大后，一旦有声音信号，就会如同电话接收方式一样，激活扬声器工作。由演播室的拍摄图像转换为电磁通信波，到电磁通信波再转换为重现于你电视机上的可视图像，这一转换过程稍微有点儿复杂。

Unit 8 Cloud Computing

Learning Objectives

After completing this unit, you will be able to do the following:

1. Grasp the main idea and the structure of the text;
2. Master the key language points and grammatical structure in the text;
3. Understand different aspects in which the cloud computing is applied in our life;
4. Conduct a series of reading, listening, speaking and writing activities related to the theme of the unit.

Outline

The following are the main sections of this unit.

1. Warm-up Activity;
2. Text A;
3. Situational Conversation;
4. Reading Comprehension;
5. Translation Skills;
6. Exercises.

Aviation Terms

In this unit, you will learn some technical terms in cloud computing listed below:

- memory;
- gigabyte;
- physical backup.

Vocabulary

The listed below are some words appearing in this unit that you should make part of your vocabulary:

- command;
- server;
- virus;
- hacker;
- hard drive;
- back up;
- decent;
- dust off.

Looking Ahead

Cloud computing

The concept of cloud computing was put forward by Google, which is a beautiful network applications mode. Based on the increase in Internet-related services and delivery models, it provides dynamic, easy to expand and virtualized resources usually through the Internet. This service can be related to IT, software and Internet; other services can be supported too. It means that computing power can be circulation as a commodity. As we expect, more and more enterprise put this new technology into practical use, for instance, webmail, G-mail and apple store use it for computing and storage. With the technology development, cloud computing has given full play to its roles. Industry, Newcomers and Media with computer technology make network cheaper, faster, convenient and better control.

Four Characteristics of Cloud Computing

Cloud computing has four distinct characteristics:

First, cloud computing provides the most reliable and secure data storage center. Second, the cloud computing on the client devises requires a minimum disk space, and the using of which is most convenient. And the third and fourth characters are its enhanced computing power and unlimited storage capacity.

Warm-up Activity

Could you provide some suggestions to strengthen the information security of our society?

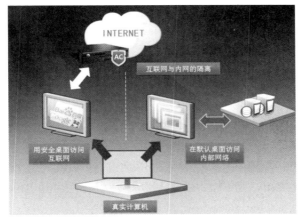

Security isolation

Remote terminal

Part one: Text A Cloud Computing Applications in Life

Nowadays, a popular website game in China and the World—"Stealing food". From Tencent "QQ farms" to the "Facebook farm community"; from one country "Farm" to the world's " Farm". People seem to look for the joy with the computers linked by internet, through this social platform to get in touch with people all over the world.

Nowadays, China and even around the world have also a popular computing—Cloud Computing. From Google, to AMD; "Cloud Computing" from a country, to the world. Technical staff seem to find value in this area, looking for the joy of success. However, the general public is difficult to "Farm" and" Cloud Computing" linked together. The reason I think is that most people do not understand cloud computing, which makes people only know how to use it without knowing how it works.

In fact, the "farm" is using cloud computing technology and its working principle is very simple. Here can be divided into 3 steps:

(1) When we click on the button "harvest", our computer will send "harvested" command into the servers in "cloud".

(2) Servers according to your instructions which are sent, through its own processing power, over write the data, and save on the server's hard disk.

(3) The servers will send back the modified data to your computers, then the computers change the display according to the data.

Our computer did not participate in the calculation of the main aspects. The task of it is to send data and receive data, which is called "Cloud Computing."

Cloud computing is an Internet-based formula, in this way, to share resources and information to on-demand hardware and software available to computers and other equipment in order to run like the whole grid.

Since cloud computing is an new-emerging computing technology, there will certainly be a lot of advantages.

Cloud computing has four distinct characteristics:

First, cloud computing provides the most reliable and secure data storage center, users do not have to worry about data loss, virus attack and other problems. All your data is stored in the cloud. It has been somewhere in the cloud. And desktop computing is different from the desktop environment, hard drive crashes will destroy all your valuable data. The cloud computer crash will not affect your data storage. Because the cloud data is automatically copied, so there will never be any loss. This also means that even if your PC crashes, all your data is still in cloud, you can still find them. A few desktop users regularly back up their data world, cloud computing can maintain the security of data.

Second, the cloud computing on the client devices requires a minimum disk space, and the using of which is most convenient.

You do not need high-performance computers to run web-based cloud application. Because the application in the cloud rather than in their own personal computers run on traditional desktop personal computer software without the required processing power or disk space. Therefore, cloud computing client computer can be cheap, with a smaller hard drive, less memory, more efficient processors. In fact, the client in this case do not even need CD or DVD drive, because it doesn't need to load any software programs and save any documents.

The third is the enhanced computing power. This is obvious. When you connect a cloud computing system, the whole of you will have discretionary power. You are no longer limited to a single computer which can do things. Using thousands of computers and servers, you can perform for supercomputing type of task. In other words, in the clouds, you may want to try greater tasks in the desktop.

The fourth is the unlimited storage capacity. Similarly, the clouds provide almost unlimited storage capacity. Imagine your desktop or notebook computer is about to run out of storage space. And the cloud can be used in hundreds of PB (100 million gigabytes) capacity compared to your computer's insignificant 200 GB hard drive-based capacity. If you need to save something, you have enough space.

Everything has two sides, so does cloud calculation. Apart from the four notable advantages, it also has some disadvantages.

1. Stored data may be unsafe

Customers making cloud computing work will reduce the degree of data security. File access is also likely to be ultra vires, and will be more dependent on the network; if hackers or criminals hacking use the cloud server, it threats other people's information security.

2. Requires sustained Internet connection

If you can not connect to the Internet, then, frankly, cloud computing is impossible. Because you use the Internet to connect to your application and documentation, if you do not have Internet connection, then you can not access anything, even if the document is yours. Internet connection failure means malfunction of cloud calculation, so Internet access is the decisive factor which may affect the use of cloud computing.

3. May be slow

Even in a fast web connection on the given application, on the desktop sometimes similar software programs are being used. This is because all the procedures, from the interface to work with your files, must be in your computer and the cloud passed back and forth between computers. If the cloud is just at that moment the server to be backed up, or if the Internet is slow, then you will not get you used to using desktop applications with the same instantaneous access.

4. If "Cloud" lost your data, then you are of "blackmail"

In theory, the data stored in the cloud will be copied to multiple machines, so to provide extraordinary security. But what if your data is really gone, you do not have any physical or local backup. (Unless you plan to download the "Cloud Files" in your desktop, in fact, very few users do so.) In short, if the "cloud" have something wrong, relying on the cloud will make your exposure to risk.

Words and Expressions

command	[kə'mɑːnd]	n. 指挥，控制；命令
server	['sɜːvə]	n.［计］服务器
overwrite	[əʊvə'raɪt]	vt. 重写，覆盖
virus	['vaɪrəs]	n.［病毒］病毒
desktop	['desktɒp]	n. 桌面；台式机
high-performance	[ˌhaɪpə'fɔːməns]	adj. 高性能的；高效能的
memory	['mem(ə)rɪ]	n.［计］内存
load	[ləʊd]	vt. 下载
discretionary	[dɪ'skreʃ(ə)n(ə)rɪ]	adj. 任意的；自由决定的
supercomputing		n.［计］超级计算
gigabytes	['gɪgəbaɪt]	n. 十亿字节；十亿位组
ultra vires		超越权限（源自于拉丁语）
hacker	['hækə]	n. 电脑黑客
offline	[ɒf'laɪn]	adj. 脱机的；离线的
social platform		社交平台
processing power		处理能力
hard disk		硬盘
data storage center		数据存储中心
hard drive		硬件驱动
hard disk crash		硬盘崩溃
back up		（资料）备份
physical backup		物理备份
local backup		本地备份

Exercise 1: Special terms.

1. high-performance ＿＿＿＿＿＿＿＿＿＿
2. desktop computer ＿＿＿＿＿＿＿＿＿＿
3. memory ＿＿＿＿＿＿＿＿＿＿
4. 笔记本电脑 ＿＿＿＿＿＿＿＿＿＿
5. 高性价比的 ＿＿＿＿＿＿＿＿＿＿
6. 只读内存 ＿＿＿＿＿＿＿＿＿＿

Exercise 2: Answer the following questions.

1. Why is it difficult for the public to link "Farm" and "cloud computing" together?

2. What's the working principle of cloud computing in the "Farm"?

3. What are the advantages of cloud computing?

4. What are the disadvantages of cloud computing?

Exercise 3: Define the following term with information from Text A.

cloud computing

📑 Part Two: Situational Conversation

(There are two colleagues discussing about a computer.)

A: Is your computer fast enough?

B: Oh my gosh, don't even ask me that question! My computer is unbelievably slow. It's the bane of my existence!

A: You're exaggerating! It can't be that bad. Didn't human resources just issue you a new computer last month? If it's new, the processor should be pretty decent. It shouldn't be that slow.

B: It's slower than my dead grandmother! The computer they gave me was new to me, but it was already 5 years old. They just dusted off some relic from the back storage room and presented it to me as new!

A: Is that true? Well, that's what you get for complaining that our company's technology is out of date. They probably did it on purpose just to spite you. If it is five years old, it's probably slower than the one you had before.... How much RAM does it have?

B: I'm not sure, but I know my computer definitely does not have enough memory. The hard drive is cluttered with so much useless software. It might be a little faster after I delete all the programs I don't often use, but what I really wish for is something state of the art. ... high speed, built-in wireless internet, enough gigabytes to satisfy...

A: Keep dreaming.

Notes:

1. It's the bane of my existence!
这是我的灾星（bane：灾星，祸根，惹人烦心的事）。

2. the processor should be pretty decent
计算机运行处理得应该相当好（decent：相称的，合宜的；相当好的）。

3. It's slower than my dead grandmother
比我死去的祖母都慢。

81

4. dust off：除去灰尘

5. RAM Random Access Memory

随机存取存贮器

6. The hard drive is cluttered with so much useless software.

硬盘里充斥着这么多无用的软件（clutter：乱糟糟地堆满）。

Exercise 1: Sentence patterns.

1. My computer is unbelievably slow.

2. Didn't human resources just issue you a new computer last month?

3. If it's new, the processor should be pretty decent.

4. I know my computer definitely does not have enough memory.

5. The hard drive is cluttered with so much useless software.

6. It might be a little faster after I delete all the programs I don't often use.

Exercise 2: Complete the following dialogue in English.

(Two friends are discussing about plasmas and MP4.)

A: Did you see the latest flat-screen plasmas that have come out? It's amazing what they come up with these days….

B: Amazing, yeah. But those suckers are pretty expensive. I heard they cost next to 600 dollars a piece.

A: Yeah, I heard that too. _____. （尽管一些新出的技术还是买得起的）

B: Like what?

A:_____. （新出的 IBM MP4 播放器定价就很合理） You can get a nice little MP4 with about 3 gigs of space for a little over a hundred bucks. _____, （我认为这是各非常合适的价格）don't you?

B: I guess so, _____. （但是屏幕小、像素不高使得它难以产生有价值的图像）You can listen to music fine, _____. （但是如果你想播放多媒体软件，几乎看不了）With the current technology limitations, I think portable MP4 players are a rip-off.

A: _____, （你想等到下一代出来之后再买）huh?

B: You got it.

Part Three: Reading Comprehension

No document is safe any more. Faking once the domain of skilled deceivers that used expensive engraving（雕刻）and printing equipment, has gone mainstream since the price of desktop-publishing systems has dropped. In ancient times, faking was a hanging offence. Today, desktop counterfeiters have little reason to worry about prison, because the systems they use are universal and there is no means of tracing forged documents to the machine that produced them. This, however, may soon change thanks to technology development by George Chiu, an anti-faking engineer.

His approach is based on detecting imperfections in the print quality of documents. Old-school court scientist were able to trace documents to particular typewriters based on quirks（瑕疵）of the individual keys. He employs a similar approach, exploiting the fact that the rotating drums and mirrors inside a printer are imperfect pieces of engineering which leave unique patterns of banding in their products.

Although these patterns are invisible to the naked eyes, they can be detected and analyzed by computer programs, and it is these patterns that Dr. Chiu has spent the past year devising. So far, he cannot trace individual printers, but he can tell pretty reliably which make and model of printer was used to create a document.

That, however, is only the beginning. While it remains to be seen whether it will be possible to trace a counterfeit（伪造）document back to its guilty creator on the basis of manufacturing imperfections. Dr. Chiu is now working out ways to make those imperfections deliberate. He wants to modify the printing process so that unique, invisible signatures can be incorporated into each machine produced which would make any document traceable.

Ironically, it was after years of collaborating with printing companies to reduce banding and thus increase the quality of prints, that he came up with the idea of introducing artificial banding that could encode identification information into a document. Using the banding patterns of printers to secure documents would be both cheap to implement and hard, if not impossible, for those without specialist knowledge and hardware to hide out.

Not surprisingly, the American Secret Service is monitoring the progress of this research very closely, and is providing guidelines to help Dr. Chiu to travel in what the service thinks is the right direction, which is fine for catching criminals. But how the legal users of printers will react to Big Brother being able to track any document back to its source remains to be seen.

1. By saying no document is safe any more, the author probably means _____.

A) cheap printers make it possible for anyone to forge documents

B) the American Secret Service will be able to trace any document

C) every printed document will be secretly marked out through high-tech

D) counterfeiters have more advanced technology to use

2. The core of both old and new ways of anti-counterfeiting is _____.

   A) the quirks of the keys of the typewriters

   B) the drums and mirrors of the printers

   C) the subtle flaws of printing devices

   D) the special skills of the experts

3. According to Dr. Chiu, what makes it possible to track down any paper work is to _____.

   A) make all the imperfections of machines deliberately

   B) find out the guilty creator of the counterfeit document

   C) allow more imperfectly designed printers to be sold

   D) mark all the printers in a special and secret way

4. The advantage of using banding patterns to trace documents is that it _____.

   A) is economically affordable and technically practical

   B) strengthens the collaboration among printing companies

   C) makes printers cheaper and harder to be taken around

   D) is able to identify even the most specialized criminals

5. It can be inferred from the last paragraph that _____.

   A) Dr. Chiu will be remembered for his special contribution

   B) the tracking of all documents might result in controversies

   C) American Secret Service is funding Dr. Chiu's research

   D) printer manufacturers are reluctant to implement deliberate banding

Part Four: Translation Skills

包孕法

这种方法多用于英译汉。所谓包孕是指在把英语长句译成汉语时，把英语后置成分按照汉语的正常语序放在中心词之前，使修饰成分在汉语句中形成前置包孕。但修饰成分不宜过长，否则会造成拖沓或成汉语句子成分在连接上的纠葛。如：

例 1： You are the representative of a country and of a continent to which China feels particularly close.

翻译：进行您是一位来自于使中国倍感亲切的国家和大洲的代表。

详解：翻译有时需要全面考虑从句的意思，比如例 1 直译就是"您是一位来自于使中国倍感亲切的国家的代表"，意思并不完整。

例 2： What brings us together is that we have common interests which transcend those differences.

翻译：使我们走到一起的，是我们有超越这些分歧的共同利益。

详解：进行翻译有时需要全面考虑从句的意思，比如该例如果直译就是"使我们走到一起的，是共同利益"，那么这个定语"是我们有超越这些分歧的"就没有翻译出来，意思并不完整。

Translation Exercise 1

对于公司和个人用户，计算机网络都有众多的用途。对公司而言，把个人计算机通过共享服务器连成网络，既有灵活性，又有良好的性价比。对个人而言，通过网络可以分享各种信息及娱乐资源。

一般来说，计算机网络可以分为 LAN（局域网）、MAN（城域网）、WAN（广域网）以及互联网等类型，每一种类型的网络都有各自的特点、技术、速度以及应用范围。局域网可以覆盖一座高楼，城域网可以覆盖整个市区，广域网则能够覆盖一个国家或大洲。

Translation Exercise 2

国际电联正在全球范围内进行一项公众调查，以评估消费者对于网络交易的信任程度和网络安全措施的意识程度。本次调查收集的数据将用于加强全球（尤其是发展中国家）对于网络安全重要性的认识，并帮助决策者评估网络的信任级别，以制定各个国家或企业的网络优先策略。

Key to the Exercises

Key of Unit 1

Part One

Exercise 1

1. 电气和机械
2. 电机工程
3. 机械工程
4. electric signal
5. mechanical linkage
6. a single electrical component

Exercise 2

1. In engineering, electromechanics combines electrical and mechanical processes and procedures drawn from electrical engineering and mechanical engineering.

2. Strictly speaking, a manually operated switch is an electromechanical component, but the term is usually understood to refer to devices which involve an electrical signal to create mechanical movement, or mechanical movement to create an electric signal.

3. The Strowger switch, the Panel switch, and similar ones were widely used in early automated telephone exchanges.

4. As in 2010, approximately 16,400 people work as electro-mechanical technicians in the US.

Exercise 3

① Electromechanic combines electrical and mechanical processes and procedures drawn from electrical engineering and mechanical engineering. One thing, made of silicon mixed with other chemical, whose resistance value is between the conductor and insulator.

② Relays originated with telegraphy as electromechanical devices used to regenerate telegraph signals. The sliced crystalline silicon which is produced in the process of purifying silicon.

③ Microcontroller circuits containing ultimately a few million transistors, and a program to carry out the same task through logic, with electromechanical components only where moving parts, such as mechanical electric actuators, are a requirement.

Part Two

Exercise 1

1. It's very nice of you to invite me.

你真是太好了，请我来做客。

2. Chinese dishes are exquisitely prepared, delicious, and very palatable. They are vey good in colour, flavour and taste.

中国菜做得很精细，色、香、味俱全。

3. Could you tell me the different features of Chinese food?

请给我讲讲中国菜的不同特色好吗？

4. It's an informal dinner.

这是一次家常便饭。

5. I'm quite full.

我已经够饱了。

Exercise 2

Complete the following dialogue in English.

1. We would be happy to give you help.

2. Would you please send my thanks to Mr. Zhang and other friends?

3. Of course, I will. Well, it's time to say goodbye, the plane is about to take off, I hope you will have the opportunity to come to the United States in the future.

4. Thank you. If I had the chance, I would go. Goodbye, May you be safe throughout the journey!

Part Three

1. B 2. D 3. C 4. D 5. C

Part Four

Translation Exercise 1

Devices which carry out electrical operations by using moving parts are known as electromechanical. Strictly speaking, a manually operated switch is an electromechanical component, but the term is usually understood to refer to devices which involve an electrical signal to create mechanical movement, or mechanical movement to create an electric signal, often involving electromagnetic principles such as in relays, which allow a voltage or current to control other, oftentimes isolated circuit voltage or current by mechanically switching sets of contacts, and solenoids, by which the voltage can actuate a moving linkage as in solenoid valves. Piezoelectric devices are electromechanical, but do not use electromagnetic principles. Piezoelectric devices can create sound or vibration from an electrical signal or create an electrical signal from sound or mechanical vibration.

Translation Exercise 2

Tenderers shall list and offer separately after the item tender offer sheet the cost per person per day for Buyer's personnel to attend designing communication meetings and factory inspection outside China and to receive trainings from Seller outside China. Expenses of the above three items will serve as references for buyer when the contract is signed, but will not be included into the tender sum. It is to make it convenient for buyer to compare and choose from the combinations of the final contract price. It does not restrict the rights of buyers to choose one quotation or the combination of several quotations to sign the contract.

Key of Unit 2

Part One

Exercise 1

1. 航空航天工程
2. 冶金工程
3. 土木工程
4. electrical engineering
5. manufacturing engineering
6. Chemical engineering

Exercise 2

1. Mechanical engineering is the discipline that applies the principles of engineering, physics, and materials science for the design, analysis, manufacturing, and maintenance of mechanical systems.

2. The engineering field requires an understanding of core concepts including mechanics, kinematics, thermodynamics, materials science, structural analysis, and electricity.

3. Mechanical engineering emerged as a field during the industrial revolution in Europe in the 18th century.

4. Subdisciplines of mechanics include

Statics, the study of non-moving bodies under known loads, how forces affect static bodies

Dynamics (or kinetics), the study of how forces affect moving bodies

Mechanics of materials, the study of how different materials deform under various types of stress

Fluid mechanics, the study of how fluids react to forces

Kinematics, the study of the motion of bodies (objects) and systems (groups of objects), while ignoring the forces that cause the motion. Kinematics is often used in the design and analysis of mechanisms.

Continuum mechanics, a method of applying mechanics that assumes that objects are continuous (rather than discrete)

Exercise 3

Mechanical engineering

Mechanical engineering is a discipline that applies the principles of engineering, physics, and materials science for the design, analysis, manufacturing, and maintenance of mechanical systems.

Mechatronics

Mechatronics is the combination of mechanics and electronics. It is an interdisciplinary branch of mechanical engineering, electrical engineering and software engineering and it concerned with integrating electrical and mechanical engineering to create hybrid systems.

Robotics

Robotics is the application of mechatronics to create robots, which are often used in industry to perform tasks that are dangerous, unpleasant, or repetitive.

Part Two

Exercise 1

1. Will you do me a favor to lift the heavy box? 你能我提下这个重箱子吗？

2. What is the problem/What is wrong/What is the matter with the machine? 机器出了什么问题？

3. If you were in my position, you'd have done the same, I am sure. 我相信如果你处在我的位置上，也会这么做。

4. Personally, I think you'd done the same, I am sure. 我个人觉得你应该努力学会如何更好地识图。

5. Only in t his way, can you do your work well. 只有这样，才能做好你的工作。

6. I will follow your advice and redouble my efforts to make up for knowledge of drawing? 我会听取您的建议，加倍努力补习关于图纸方面的知识。

7. If you have any questions, don't hesitate to ask. 有什么问题只管问。

8. I am honored to do something useful for you. 能为你做些有用的事情我深感荣幸。

Exercise 2

A: Excuse me ! Can you answer me some questions about building drawings?

B: Certainly. What do you want to know about drawings?

（当然可以。关于图纸你想了解什么？）

A:First, I want to know a more scientific definition of drawings.

B:The so-called drawings are the designs drawn minutely and accurately by the method of orthogonal projection in accordance with the relevant regulations for the proposed buildings in the content of the internal and external shape and size, the structure, construction,decoration and equipment of different parts, and so on.

（所谓的图纸就是将内外形状和大小以及各部分的结构、构造、装饰、设备等内容，按照有关规范规定，用正投影方法，详细准确地画出的图样。）

A: So drawings are very important to the construction works, aren't they?

B: You are right. Drawings have always been regarded as the norm and the guide of our construction works.

（你说的没错。图纸一直被看作施工的指南和准则。）

A: How many parts are there in a complete set of mechanical drawings?

B: According to the professional content and functions, a complete set of mechanical drawings generally include the catalogue, the general design specification, the building construction drawing, the structure and construction.

（一套完整的机械制图，根据专业内容或者作用不同，一般包括图纸目录、设计总说明、建筑施工图、结构施工图以及设备施工图。）

A: I know. They are mainly used in manual drawings, aren't they?

B: Right. At present, the mechanical drawings are mainly finished by the computer A utoCAD software, the advantages of which are self-evident in contrast to the manual drawing. （没错，现在的机械制图大多是通过计算机 AutoCAD 软件绘制出来的。相对于手工绘图，其优点是不言而喻的。）

A: What is AutoCAD, I want to know?

B: AutoCAD is an automated computer-aided design software developed by the America Autodesk Corporation for two-dimensional and three-dimensioned design and graphing.

（AutoCAD 是美国 Autodesk 公司开发的用于二维及三维设计、绘图的自动计算机辅助软件）

A: Would you mind teaching me to make drawings by AutoCAD something when you are free?

B:Of course not. If necessary, you can give me a call, and I will do my best to help you. （当然不介意，如有需要，只管给我打电话，我一定会不遗余力地帮你。）

Part Three

1. C   2. A   3. D   4. C   5. D

Part 4

Translation Exercise 1

Mechanical engineering emerged as a field during the industrial revolution in Europe in the 18th century; however, its development can be traced back several thousand years around the world. Mechanical engineering science emerged in the 19th century as a result of developments in the field of physics. The field has continually evolved to incorporate advancements in technology, and mechanical engineers today are pursuing developments in such fields as composites, mechatronics, and nanotechnology. Mechanical engineering overlaps with aerospace engineering, metallurgical engineering, civil engineering, electrical engineering, manufacturing engineering, chemical engineering, and other engineering disciplines to varying amounts. Mechanical engineers may also work in the field of biomedical engineering, specifically with biomechanics, transport phenomena, biomechatronics, bionanotechnology, and modelling of biological systems.

Translation Exercise 2

If the alarm is not specified, check whether the E1 cable connecting the interface is loosely connected. Reconnect them and check whether the alarm is removed. If the alarm still exists, perform equipment loopback and line loopback of the E1 card to confirm the failed card.

Key of Unit 3

Part One

Exercise 1

1. 机器人
2. 数控（技术）
3. 智能机器人
4. programmable
5. built-in control system
6. remote control

Exercise 2

1. While with NC only a point, namely the endpoint of the cutter, is controlled in the space, with robots both the endpoint and orientation are manipulated.

2. The Robot Institute of America (RIA) defines the industrial robot as "a reprogrammable multi-functional manipulator designed to move material, parts, tools, or other specialized devices through variable programmed motions for the performance of a variety of tasks."

3. It can be a welding head, a spray gun, a machining tool, or a gripper containing on-off jaws, depending upon the specific application of the robot.

4. No.

Exercise 3

1. a programmable manipulator, equipped with the end-effector, capable of carrying out factory work and stand-alone operation

2. a robot that can see, hear, and touch

3. mechanical mechanism

Part Two

Exercise 1

1, 4, 3, 2

Exercise 2

3, 4, 1, 5, 2

Part Three

1. D 2. B 3. D 4. C 5. B

Part Four

Translation Exercise 1

Industrial robots are beginning now to revolutionize industry. These robots do not look or behave like human beings, but they do the work of humans. Robots are particularly useful in a wide variety of applications, such as material handling, spray painting, spot welding, arch welding, inspection, and assembly. Current research efforts focus on creating a "smart" robot that can "see", "hear, "and "touch" and consequently make decisions.

Translation Exercise 2

Within warranty period, the consumer should show the valid purchase invoice of the appliance and other relevant guarantee documents defined by the manufacturer when asking for free of charge repair. Consumer should keep the warranty card and the purchase invoice properly and show them together when asking for free of charge repair. The warranty period starts from the invoice date. If the customer loses the invoice, it starts from the manufacturing date.

新编机电英语

Key of Unit 4

Part One

Exercise 1

1. 机械化
2. 自动化
3. 大规模生产的产品
4. self-regulate
5. hydraulic
6. load and unload

Exercise 2

1. No.

2. An automatic mechanism is a mechanism which has a capacity for self-regulate; that is, it can regulate or control the system or process without the need for constant human attention or adjustment.

3. Feedback is based upon an automatic self-regulating system and any deviation in the system from desired conditions can be detected, measured, reported and corrected by virtue of it.

4. It has been highly mechanized ever since the 1920s.

Exercise 3

1. Automation is a process which is concerned with the operation and control of a complete producing unit.

2. It refers to machine which may be as simple as a convey or belt to another.

3. It refers to a process which requires the employment of human labor to control each machine as well as to load and unload materials and transfer them from one place to another.

Part Two

Exercise 1

3, 1, 4, 2

Exercise 2

Omitted

Part Three

1. D 2. C 3. B 4. C 5. A

Part Four

Translation Exercise 1

Processes of mechanization have been developing and becoming more complex ever since the beginning of the Industrial Revolution at the end of the 18th century. The current development of automatic processes is, however, different from that of the old ones. The "automation" of the 20th century is distinct from the mechanization of the 18th and 19th centuries in as much as mechanization was applied to individual operations, whereas "automation" is concerned with the operation and control of a complete producing unit. And in many, though not all, instances the element of control is so great that mechanization displaces muscle, "automation" displaces brain as well.

Translation Exercise 2

High definition: this unit adopts MPEG2 coding format and brings the horizontal resolution over 500 lines.

Time search: it can quickly search a specific part on a disc, especially agreeable for playing action movies.

Content display: TFT LCD and Chinese/ English OSD make disc contents clearer.

Key of Unit 5

Part One

Exercise1

1. 机电一体化 , 机械电子学
2. 力学，机械学
3. 电子学
4. interdisciplinary
5. optoelectronic
6. actuator

Exercise 2

1. No.

2. A Japanese engineer from Yasukawa Electric Company invented the term Mechatronics in 1969.

3. In 1996.

4. Modelling and design, system integration, actuators and sensors, intelligent control, robotics, manufacturing, motion control, vibration and noise control,microelectronic devices and optoelectronics systems, automotive systems and other applications.

Exercise 3

1. The synergistic integration of mechanical engineering with electronics and intelligent computer control in the design and manufacture of industrial products and processes.

2. Until the early 1980s, mechatronics meant a mechanism that is electrified.

3. In the mid-1980s, mechatronics came to mean engineering that is the boundary between mechanics and electronics.

Part Two

Exercise 1

3, 1, 2

Omitted

Part Three

1. B 2. D 3. A 4. C 5. D

Part Four

Translation Exercise 1

Mechatronics is nothing new; it is simply the application of the latest techniques in precision mechanical engineering, control theory, computer science, and electronics to the design process to create more functional and adaptable products. This, of course, is something that many forward-thinking designers and engineers have been working on for years.

Translation Exercise 2

Troubles and troubleshooting: if the unit makes no response when power button is pressed, please check if ① the power supply is already turned on; ② the correct input is chosen; ③ the volume is turned to its minimum level; ④ the speakers are correctly connected; and ⑤ the main unit is set at the "mute" mode.

Key of Unit 6

Part One

Exercise 1

1. 接口
2. 产生电流
3. 电子信号
4. conjunction box
5. wind power generation
6. digital signal

Exercise 2

1. It is made of silicon.

2. Common silicon is black brown, while pure silicon is dark brown.

3. Conventional distillation process industrially, you can get more pure silicon, then the final regional refining method, silicon purification to about every million silicon atoms have an impurity atom or purer.

4. The reduction may be made of silicon solar cells which can produce electricity, but the current efficiency is not very good. Silicon compounds can also be made into a lot of very useful things, such as silicon carbide, which is a very hard material; another example glass, it is very commonly used material.

Exercise 3

1. One thing, made of silicon mixed with other chemical, whose resistance value is between the conductor and insulator.

2. The sliced crystalline silicon which is produced in the process of purifying silicon.

3. One thing made of silicon with the feature of amplifying the electronic signals.

Part Two

Exercise 1

1. I am planning on sending a bulk email to all the users on our company's server. 我计划向我们公司服务器上的所有用户发群邮件。

2. That's a time-saving way to get the word out. 这是一个节省时间的宣传办法。

3. Most of us just delete the bulk emails without even reading them. 大多数人可能读都不读就把群邮件删了。

4. You could send the meeting brief to the managers by email, and ask them to forward it to their subordinates. 你可以先把会议提要通过邮件发给各部门经理，再让经理转发给他们的下属。

5. I can just put the meeting minutes on the email as an attachment, then forward it along to the managers. 我可以把会议纪要上传至附件中，一起发送给各部门经理。

6. I know next to nothing about the new email program. 我对新的邮件程序几乎一无所知。

7. but I still haven't figured it out 但我仍然没搞清楚。

Exercise 2

A: Let's correspond by email. <u>I think we should definitely keep in touch.</u>（我认为我们应该保持联系）There are a lot of future opportunities to work together.

B: I agree. <u>I am especially interested in the Aluminum project you mentioned.</u>（我对你所说的 Aluminum 项目尤其感兴趣）<u>Do you think you can email the details to me?</u>（你能把细节发送给我吗？）

A: Sure! <u>I'll send it along to you as soon as I get back to the office.</u>（我一回到办公室就一并发给您）I did get your card, didn't I?

B: Oh, I almost forget! Here it is.

A: Thank you. Is all the information on here current?

B: Let's see… Yes, but there is only my work email address. <u>I'll give you my personal address too. Sometimes if the attachment is too large, my work email will reject it.</u>（有时候如果附件过大，我的工作邮箱就会拒收）. <u>If your attachment is more than 100KB, go ahead and send it to my personal email address instead.</u>（如果你的附件超过 100KB，请发至我的个人邮箱）

A: It'll be alright. <u>I can send a compressed file.</u>（我可以发一个压缩文件）Does your computer have the software to unzip files?

B: I do, but unzipping files doesn't work out so well. <u>Last time I tried to decompress a file, my whole system crashed.</u>（上一次我试着解压文件，整个电脑系统都崩溃了）If it is a large file, it would probably work better to send it to my personal email. It's better to be safe than sorry.

A: No problem, I will make sure to email the information to your personal email address first thing.

Part Three

1. A 2. D 3. C 4. B 5. C

Part Four

Translation Exercise 1

Thank you for using this digital product of our company. In order to let you experience the product easily, the detailed instruction is provided by which you can find the product's introduction, usage and other information. Before using this product, please read the manual carefully, so that you can use it correctly. In case of any printing or translation errors, we apologize for the inconvenience. As for the change of content, we are sorry for no further notice.

This product is a well-designed portable multimedia mini acoustics which applies to household, outdoor travel, office and other places. It offers you a chance to indulge yourself in music at any time and in any place, and provides perfect sound service to your computers, digital media players, mobile phones and other audio-visual products.

Translation Exercise 2

Microcomputer system of this product can automatically identify exterior equipment. After startup and entering into standby mode, insert U-disk, TF-card or audio signal lines for automatic identification and be played according to the principle of the later coming first. For more details, please reference to the 4th item "Definition of Button and Jack".

One end of wiring can be inserted in DC 5 V, while another end with USB jack can be inserted in USB jack of PC or another charging with standard 5 V 500 mA jack, twinkle under charging and no twinkle after fully charged.

Key of Unit 7

Part One

Exercise 1

1. 无线网络

2. 无线接入

3. 访问速度

4. wireless communication

5. wireless broadband access

6. smart refrigerator

Exercise 2

1. Mobile communication is moving from simple voice to rich media, where we use more of our senses to intensify our experiences.

2. Compared with 2G, 3G will significantly boost network capacity—so operators will be able to support more users, as well as offer more sophisticated services.

3. The services that 3G enables will help us to manage our personal information, simplify tasks such as grocery shopping, make better use of our time and offer services that are just fun to use.

4. 3G will not just support the needs of business people who travel a lot, but will also help new, flexible working practices, such as home-working and remote access to corporate networks outside traditional working hours.

Exercise 3

1. IP is packet-based, which in simple terms, means users can be "online" at all times, without having to pay until we actually send or receive data.

2. 3G brings together high-speed radio access and IP-based services into one powerful environment, and introduces wideband radio communications, with access speeds of up to 2 Mbit/s.

Part Two

Exercise 1

1. Our company is researching and developing a new product!

我们公司正在研发新产品！

2. They fought in R&D in the first line.

他们战斗在研发一线。

3. The competitors knew you can win.

竞争对手知道你们能赢。

4. Next, make product test again and again until the product success rate reaches 100%.

接下来一遍一遍做测试直到产品合格率达到百分之百。

Exercise 2

(Two colleagues are discussing about a project leader.)

A: Hi, good afternoon. How are you?

B: It's OK. I heard you are going to work on a project recently . （我听说你最近要做一个项目）

A: Yes. But we haven't a project leader. So, we need to choose one.

B: What do you need of the ideal qualities of project leader? （你需要项目领导人具备哪些良好的素质）

A: Basically, we need someone who can coordinate the members of the team. （我们需要一个人能够与团队的其他成员协同一致）

B: It's very nice. In my opinion, leader can bring us a harmonious environment. What else?

A: Yes, but this person should also be someone who can motivate and lead people. （但是这个人也应该能够激励和领导大家）

B: Yes. It is good for team, and it is very important to the progress. What's your opinion?

A: A good project leader must represent the group's interests in the larger organization. （一个好的项目领导人必须能在更大的组织面前代表他们团队的利益） They should be a sort of champion for the project.

B: I agree it. That's great and then what?

A: At the same time. We need a very responsible person. （我们需要一个有责任感的人）

B: Yes. You're right. It's the project leader who has to keep the project on schedule and control the budget. （正是项目领导人把项目提上日程并且按照预算来做的）

A: Thank you!

B: You're welcome. See you.

Part Three

1. D   2. D   3. B   4. A   5. C

Part Four

Translation Exercise 1

ButtonPhone has a few simple buttons, Screen home key: Can make you keep returned to the main screen interface. Side key: the switch key, can switch between the menu key(back key) and volume key. Power button: When you press and hold this button you can power off or power on the phone; when your phone is opened, press this button to snooze or awake the display screen. If you currently do not use the phone, you can lock it to turn off the display screen to save battery power. If the phone is locked, the touch screen has no effect. Phone can still receive calls, receive text messages and receive other updates. You can also listen to music and adjust the volume, and use the center button on the headset (or the equivalent button on the Bluetooth headset) to play or pause a song, or answer or end a call. By default, if you do not touch the screen for one minute, the phone will lock.

Lock: Press the Power button.

Unlock: Press the Home button or Power button, then drag the slider.

Switch off: Press the Power button for a few seconds, until the phone options.

Power off: Select power off, the phone will close the applications into shutdown.

Switch on: Press the Power button until the power on logo.

Translation Exercise 2

When the energy waves from broadcasting stations sweep across the antenna of your radio or TV receive, a feeble, fluctuating current of electricity is generated in the antenna. This current is electronically separated from all other currents that are simultaneously being induced in the same antenna by other energy waves. The selected current (as a specific radio frequency or TV channel) is amplified and, in the case of an audio signal, actuates a speaker in the same manner as the telephone receiver. The conversion of a camera image to electromagnetic communications waves in the TV studio and back to a visual rendition on your TV receiver screen is somewhat more complex.

Key of Unit 8

Par One

Exercise 1

1. 高性能的
2. 台式电脑
3. 内存
4. laptop computer
5. highly cost-effective
6. read-only memory

Exercise 2

1. Because most people do not understand cloud computing.

2. The working principle can be divided into 3 steps.

(1) When we click on the button"harvest", our computer will send "harvested" command into the servers in "cloud".

(2) Servers according to your instructions which is sended, through its own processing power, overwriting the data, and save on the server's hard disk.

(3) The servers will send back the modified data to your computers, then the computers, change the display according to the data.

3. There are four advantages of cloud computing. First, cloud computing provides the most reliable and secure data storage center. Secondly, the cloud computing on the client devices require a minimum disk space, and the using of which is most convenient. Thirdly, it can enhanced computing power. Fourthly, it can provide unlimited storage capacity

4. There are four disadvantages. Firstly, stored data may be unsafe. Secondly, it requires sustained internet connection. Thirdly, the speed of cloud computing may be slow. Fourthly, if "Cloud" lost your data, then you are of "blackmail".

Exercise 3

Cloud computing is an Internet-based formula, through which computer dose not participate in the calculation of the main aspects, only has to send data and receive data. Shared resources and information to on-demand hardware, and software available to computers and other equipment run like the whole grid.

Part Two

Exercise 1

1. My computer is unbelievably slow.

我的电脑慢到不行了。

2. Didn't human resources just issue you a new computer last month?

人力资源部上个月不是给你一台新电脑吗？

3. If it's new, the processor should be pretty decent.

如果电脑是新的，运行速度应该相当快。

4. I know my computer definitely does not have enough memory.

我知道我的电脑没有足够的内存了。

5. The hard drive is cluttered with so much useless software.

硬件驱动上装了太多没有用的软件。

6. It might be a little faster after I delete all the programs I don't use often.

如果我删除一些不常用的软件，电脑可能会快些。

Exercise 2

(Two friends are discussing about plasmas and MP4.)

A: Did you see the latest flat-screen plasmas that have come out? It's amazing what they come up with these days…

B: Amazing, yeah. But those suckers are pretty expensive. I heard they cost next to 600 Dollars a piece.

A: Yeah, I heard that too. Some of the new technology that comes out is pretty affordable, though. （尽管一些新出的技术还是买得起的）

B: Like what?

A: The new IBM MP4 players are pretty reasonably priced for what you get. （新出的 IBM MP4 播放器定价就很合理）You can get a nice little MP4 with about 3 gigs of space for a little over a hundred bucks. I think that's a pretty decent price, （我认为这是个非常合适的价格）don't you?

B: I guess so, but the tiny screens don't have enough pixels to make any graphics worthwhile. （但是屏幕小、像素不高使得它难以产生有价值的图像）You can listen to music fine, but if you want to play some multi-media files, you can't see much. （但是如果你想播放多媒体软件，几乎看不了）With the current technology limitations, I think portable MP4 players are a rip-off.

A: You want to wait until the next generation comes out before buying in, （你想等

到下一代出来之后再买）huh?

B: You got it.

Part Three

1. A　2. C　3. D　4. A　5. B

Part Four

Translation Exercise 1

Computer networks can be used for numerous services, both for companies and for individuals. For companies, networks of personal computers using shared computers often provide flexibility and a good price/performance ratio. For individuals, networks offer access to a variety of information and entertainment resources.

Roughly speaking, networks can be divided into LANs, MANs, WANs, and internet works, each with their own characteristics, technologies, speeds, and niches. LANs cover a building, MANs a city, and WANs a country or continent.

Translation Exercise 2

ITU is conducting a worldwide public survey to assess users' trust of online transactions and awareness of cybersecurity measures. The data collected through the survey will be used to increase global awareness of cybersecurity, particularly in developing countries, and should help decision-makers in assessing the cyberspace "trust" level with a view to reviewing national and corporate strategies and priorities.